DATE DUE

Engine Testing

Engine Testing
Theory and practice

Second edition

Michael Plint and Anthony Martyr

Published on behalf of:
Society of Automotive Engineers, Inc.
400 Commonwealth Drive
Warrendale, PA 15096-0001

Butterworth-Heinemann
Linacre House, Jordan Hill, Oxford OX2 8DP
225 Wildwood Avenue, Woburn, MA 01801-2041
A division of Reed Educational and Professional Publishing Ltd

A member of the Reed Elsevier plc group

First published 1995
Reprinted 1996 (twice), 1997 (twice)
Second edition 1999

Library of Congress Cataloguing in Publication Data
A catalogue record for this book is available from the Library of Congress

ISBN 0 7680 0314 8

Typeset by David Gregson Associates, Beccles, Suffolk
Printed and bound in Great Britain by Biddles, Guildford, Surrey

FOR EVERY TITLE THAT WE PUBLISH, BUTTERWORTH-HEINEMANN
WILL PAY FOR BTCV TO PLANT AND CARE FOR A TREE.

Contents

Appendices

Preface to the second edition

In the three years since its first publication the book has gained widespread acceptance by practising test engineers on both sides of the Atlantic. At the same time there has been considerable feedback in the form of requests for fuller treatment of various topics. This edition is an attempt to meet these needs. Major changes include a much fuller treatment of the role of the computer in the test cell, more information on test cell services of all kinds, with a new section on water quality, extended treatment of engine-dynamometer shaft design and a critical study of the problems involved in accurate performance measurement, with particular reference to the condition of the combustion air. A new chapter covers chassis dynamometers of all kinds and information regarding the rapidly evolving field of exhaust emissions has been updated and extended.

M.A.P., Wokingham
A.J.M., Inkberrow

January 1998

Preface to the first edition

It is probably true to say that the number of engine test beds in commission and the number of engine-hours of running is greater today than at any time since the invention of the internal combustion engine. The sheer number of engines in use grows continuously; all but the smallest must spend some time on the test bed. The development of electronic engine control units and the ever more elaborate management of transient behaviour represents another heavy commitment for the test department, while far more attention is given to the bed testing of complete vehicles for NVH (noise vibration and harshness) than was the case in the past.

Perhaps the most important factor is the development of emissions legislation, pioneered and driven on by the Environmental Protection Agency in the USA. This calls for a vast amount of testing of vehicle engines, in many cases of long duration, plus statutory returns to the test bed in the course of the vehicle's life.

The authors have long experience in all aspects of engine testing and several years ago became aware that much of the essentially eclectic knowledge they had amassed was not available in any readily accessible form and indeed was in danger of being lost to the current generation of young engineers, whose training and education generally lacks the practical experience of engineering test procedures that was usual in the days when technician and graduate courses had a less extensive content.

In a subject as wide as the present one it is impossible that two authors could be specialists in every aspect and help and advice from many sources must be acknowledged. Particular thanks are due to the authors' companies, Plint and Partners Ltd, and Martyr Test Technology Ltd, for their support and the provision of facilities, to Emeritus Professor Frank Wallace for a painstaking review of the entire work and to Mrs Maureen Frid for patient and accurate preparation of the manuscript. Many other companies, research establishments and individuals have contributed advice and information and the authors have endeavoured to make appropriate acknowledgements in the body of the book.

M.A.P., Wokingham
A.J.M., Inkberrow

Acknowledgements

The authors would like to acknowledge their grateful thanks to the following for their kind permission to reproduce the diagrams listed.

Figure 3.4 Reprinted with permission from 'Improving the Determination of Mass Burn Fraction', P.J. Shayler *et. al.*, SAE Paper No. 900351 © 1990 Society of Automotive Engineers, Inc.

Figure 4.4 Reprinted from CIBSE Guide C, section C4, by permission of the Chartered Institution of Building Services Engineers

Figure 4.7 Reprinted from I.H.V.E. Psychrometric Chart, by permission of the Chartered Institution of Building Services Engineers

Figure 5.7 Reprinted from 'Industrial Noise Control', Fader, by permission of John Wiley & Sons, Inc.

Figure 5.10 Reprinted from technical literature of Type TSC by courtesy of Christie and Grey Limited, UK

Figure 5.13 Reprinted from 'Industrial Noise Control', Fader, by permission of John Wiley & Sons, Inc.

Figure 5.14 Reprinted from 'Encyclopaedia of Science and Technology', Vol. 12, © 1987, by kind permission of McGraw-Hill, Inc., New York

Figure 5.15 Reprinted from 'Industrial Noise Control'; Fader, by permission of John Wiley & Sons, Inc.

Figure 5.16 Reprinted from 'Industrial Noise Control', Fader, by permission of John Wiley & Sons, Inc.

Figure 6.1 Reprinted from BS799. Extracts from British Standards are reprinted with the permission of BSI. Complete copies can be obtained by post from BSI Sales, Linford Wood, Milton Keynes, MK14 6LE

Figure 6.2 Reprinted from 'The Storage and Handling of Petroleum Liquids', Hughes and Swindells, with the kind permission of Edward Arnold publishers

Figure 6.3 Reprinted from 'Recommendations for pretreatment and cleaning of heavy fuel oil' with the kind permission of Alfa Laval Ltd

Figure 6.6 Illustration courtesy of Ricardo Test Automation Ltd

Figure 7.2 Reprinted with permission of Froude Consine, UK

Figure 7.3 Reprinted with permission of Froude Consine, UK

Figure 7.4 Reprinted with permission of Froude Consine, UK

Figure 7.7 Reprinted from Drawing no GP 10409 (Carl Shenck AG, Germany)

Figure 7.8 Reprinted from technical literature, Wichita Ltd

Figure 8.4 Reprinted from 'Practical Solution of Torsional Vibration Problems' 3rd edition, W. Ker-Wilson, 1956

Figure 8.7 Reprinted from literature, with the kind permission of British Autoguard Ltd

Figure 8.10 Reprinted from sales and technical literature, with the kind permission of Twiflex Ltd

Figure 8.11 Reprinted from sales and technical literature, with the kind permission of Twiflex Ltd

Table 8.8 Reprinted from sales and technical literature, with the kind permission of Twiflex Ltd

Figure 9.2 Illustration courtesy of Ricardo Test Automation Ltd

Figure 9.3 Reprinted from technical literature, with the kind permission of Westland Aerospace

Figure 9.4 Reprinted from Technical Documentation T 32FN, with the kind permission of Hottinger Baldwin Messtechnik GmbH, Germany

Figure 9.5 Reprinted from Technical Documentation T 32FN, with kind permission of Hottinger Baldwin Messtechnik GmbH, Germany

Figure 9.7 Reprinted from Paper ISATA, 1982, R.A. Haslett, with the kind permission of Cussons Ltd

Figure 9.11 Reprinted from Technology News with the kind permission of Petroleum Review

Pages 180–181 Reprinted from B5514. Extracts from British Standards are reprinted with the permission of BSI. Complete copies can be obtained by post from BSI Sales, Linford Wood, Milton Keynes, MK14 6LE

Figure 13.7 Reprinted from SAE 920 462 (SAE International Ltd)

Figure 14.1 Reprinted by permission of AEA Technology

Figure 14.2 Reprinted from Technological Responses to the Greenhouse Effect (Watt Committee Report No. 23) with the kind permission of the Watt Committee on Energy

Figure 14.3 Reprinted from Fundamentals of Exhaust Emissions Analysis and their Application, with the kind permission of Horiba Instruments

Figure 14.4 Reprinted from Report 2/92, with the kind permission of CONCAWE

Figure 14.5 Reprinted from Report 2/92, with the kind permission of CONCAWE

Figure 14.6 Reprinted from Automotive Handbook, Second Edition, 1986, with the kind permission of Robert Bosch GmbH

Figure 14.7 Reprinted from Report 2/92, with the kind permission of CONCAWE

Figure 14.8 Reprinted from 'Focus on Vehicle Emissions' by permission of the Council of the Institution of Mechanical Engineers from Automotive Engineer Sept/Oct 93

Figure 14.9 Reprinted from Automotive Handbook, Second Edition, 1986, with the kind permission of Robert Bosch GmbH

Figure 14.10 Reprinted from Fundamentals of Exhaust Emissions Analysis and their Application, with the kind permission of Horiba Instruments

Figure 14.11 Reprinted from Fundamentals of Exhaust Emissions Analysis and their Application, with the kind permission of Horiba Instruments

Figure 14.12 Illustration courtesy of Lucas Assembly and Test Systems

Figure 14.13 Reprinted with permission from SAE J 1280 © 1992 Society of Automotive Engineers, Inc.

Table 14.1 Reprinted from 'Encyclopaedia of Science and Technology', Vol. 16, © 1987, by kind permission of McGraw-Hill, Inc., New York

Figure 15.1 Reprinted with permission of Professor J. Wallace. (Private communication)

Figure 16.1 Reprinted with permission of Froude Consine, UK

Figure 16.2 Illustration courtesy of Ricardo Test Automation Ltd

Figure 17.1 From: Schmiertechnik und Tribologie 29, H. 3, 1982, p. 91, Vincent Verlag Hannover. (Now: Tribologie und Schmierungstechnik)

Figure 17.3 Reprinted by permission of the Council of the Institution of Mechanical Engineers from 'The effect of viscosity grade on piston ring wear', S.L. Moore, Proc. I. Mech. E C184/87

Figure 17.4 Reprinted by permission of the Council of the Institution of Mechanical Engineers from 'The effect of viscosity grade on piston ring wear', S.L. Moore, Proc. I. Mech. E C184/87

Figure 17.5 Reprinted by permission of the Council of the Institution of Mechanical Engineers from 'The effect of viscosity grade on piston ring wear', S.L. Moore, Proc. I. Mech. E C184/87

Figure 16.6 Reprinted by permission of the T&N Technology, Rugby from BP Tribology Lecture, 1990. D.A. Parker

Introduction

In 1936 Sir Harry Ricardo, whose contribution to the development of both the gasolene and diesel engine was perhaps greater than that of any other engineer, wrote:

> An engine may be regarded as a creature of infinite and dogged cussedness, but entirely lacking in sense of humour – a very fit subject for the practical joke – and the art of testing consists really in keeping our patience and inventing new practical jokes to play upon it; secure in the knowledge that we can fool it all the time[1].

These words are as true today as when they were written. The endless fascination of the internal combustion engine as a subject for research and development is a consequence of its resemblance to a living creature, capable of manifesting an almost infinite range of puzzling and sometimes inexplicable behaviour.

It follows that the engineer concerned with any aspect of engine testing, be it fundamental research, development, performance monitoring or routine production testing, must have at his fingertips a wide and ever-broadening range of knowledge and skills.

A particular problem he must face is that, while he is required to master ever more advanced experimental techniques – such areas as emissions analysis and transient testing come to mind – he cannot afford to neglect any of the more traditional aspects of the subject. Such basic matters as the mounting of the engine, coupling it to the dynamometer and leading away the exhaust gases can give rise to intractable problems, misleading results and even on occasion to disastrous accidents. More than one engineer has been killed as a result of faulty installation of engines on test beds.

The sheer range of machines covered by the general term 'internal combustion engine' broadens the range of necessary skills. At one extreme we may be concerned with an engine for a chain saw, a single cylinder of perhaps 50 c.c. capacity running at 15 000 rev/min on gasolene, with a running life of a few hours. Then we have the vast number of passenger vehicle engines, four-, six- or eight-cylinder, capacities ranging from one litre to six, expected to develop full torque over speeds ranging from perhaps 1500 rev/min up to 7000 rev/min (the upper limit rising continually, so that 10 000 rev/min is by no means unheard of), and with an expected life of perhaps 6000 hours. At the other extreme is the 'cathedral' type marine engine, a machine perhaps 10 m tall and weighing 1000 tonnes, running on the worst type of residual fuel, and expected to go on turning at 70 rev/min for more than 50 000 hours.

The purpose of this book is to bring together the information on both the theory and practice of engine testing that any engineer responsible for work of this kind must have available. It is naturally not possible, in a volume of manageable size, to give all the information that may be required in the pursuit of specialized lines of development, but a wide range of references for more advanced study bas been included.

Throughout the book *accuracy* will be a recurring theme. The purpose of engine testing is to produce information, and inaccurate information can be useless or worse. A feeling for accuracy is the most difficult and subtle of all the skills required of the test engineer. Chapter 19, dealing with this subject, is perhaps the most important in the book and the first that should be read.

Experience in the collaboration with architects and structural engineers is particularly necessary. These professions follow design conventions and even draughting practices that differ from those of the mechanical engineer. To give an example, the test cell designer may specify a strong floor on which to bolt down engines and dynamometers that has an accuracy approaching that of a surface plate. To the structural engineer this will be a startling concept, not easily achieved.

The internal combustion engine is perhaps the best mechanical device available for introducing the engineering student to the practical aspects of engineering. An engine is a comparatively complicated machine, sometimes noisy and alarming in its behaviour and capable of presenting many puzzling problems and mystifying faults. A few hours spent in the engine testing laboratory are perhaps the best possible introduction to the 'real world' of engineering, which is remote from the world of the lecture theatre and the 'computer simulation' in which, inevitably, the student spends much of his time.

While it contains some material only of interest to the practising test engineer, much of this book is equally suitable as a student text, and this purpose has been kept very much in mind by the authors.

A note of warning: the general management of engine tests

What may be regarded as 'traditional' internal combustion engines had in general very simple control systems. The spark ignition engine was fitted with a carburetter controlled by a single lever, the position of which, together with the resisting torque applied to the crankshaft, set all the parameters of engine operation. Similarly the performance of a diesel engine was dictated by the position of the fuel pump rack, either controlled directly or by a relatively simple speed governor.

In more recent times the advent of engine control units (ECUs) has entirely changed the situation. The ECU monitors many aspects of engine performance and makes continuous adjustments. The effect of this is effectively to take the control of the test conditions out of the hands of the engineer conducting the test. Factors entirely extraneous to the investigation in hand may thus come into play (one of the authors experienced a case in which a vehicle engine could not be brought up to full power because an interlock associated with the closing of the car boot had not been de-activated).

The introduction of exhaust gas recirculation (EGR) under the control of the ECU is a typical example. The only way open to the test engineer to regain control of his test is to devise means of bypassing the ECU, either mechanically or by intervention in the programming of the control unit.

A note on references and further information

It would clearly not be possible to give all the information necessary for the practice of engine testing and the design of test facilities in a book of this length.

The authors have been careful to include sufficient references and further reading to provide supplementary information. These are of two kinds:

- a selection of fundamental texts and key papers
- relevant British Standards and other standard specifications.

Reference

1. Ricardo, H.R. and Hempson, J.G.G. (1968) *The High-Speed Internal Combustion Engine*, Blackie, London.

Units and conversion factors

Throughout this book use is made of the metric system of units, variously described as:

The MKS (metre-kilogram-second) System
SI (Système International) Units

These units have the great advantage of logical consistency but the disadvantages of still, a certain degree of unfamiliarity and in some cases of inconvenient numerical values.

Fundamental Units

Mass	kilogram (kg)	$1\,\text{kg} = 2.205\,\text{lb}$
Length	metre (m)	$1\,\text{m} = 39.37\,\text{in}$
Force	newton (N)	$1\,\text{N} = 0.2248\,\text{lbf}$

Derived Units

Area	square metre (m^2)	$1\,\text{m}^2 = 10.764\,\text{ft}^2$
Volume	cubic metre (m^3), litre(1)	$1\,\text{m}^3 = 10001 = 35.3\,\text{ft}^3$
Velocity	metre per second (m/s)	$1\,\text{m/s} = 3.281\,\text{ft/s}$
Work, Energy	joule (J)	$1\,\text{J} = 1\,\text{Nm} = 0.7376\,\text{ft-lbf}$
Power	watt (W)	$1\,\text{W} = 1\,\text{J/s}$
		$1\text{ horsepower (hp)} = 745.7\,\text{W}$
Torque	newton metre	$1\,\text{Nm} = 0.7376\,\text{lbf-ft}$

The old metric unit of energy was the calorie (cal), the heat to raise the temperature of 1 gram of water by 1°C.

1 cal $= 4.1868$ J. 1 kilocalorie (kcal) $= 4.1868$ kJ.

Temperature degree Celsius °C $\qquad\qquad\qquad\qquad\qquad\qquad\qquad \theta$

Absolute temperature Kelvin K $\qquad\qquad\qquad\qquad\qquad\qquad\qquad T$

$$T = \theta + 273.15$$

Pressure Pascal (Pa) $\qquad\qquad 1\,\text{Pa} = 1\,\text{N/m}^2 = 1.450 \times 10^{-4}\,\text{lbf/in}^2$
$$1\,\text{MPa} = 10^6\,\text{Pa} = 145\,\text{lbf/in}^2$$

This unit is commonly used to denominate stress.
Throughout this book the bar is used to denominate pressures:
1 bar (bar) $= 10^5\,\text{Pa} = 14.5\,\text{lbf/in}^3$

Standard test conditions for i.c. engines as defined in BS 5514/ISO 3046[1] specify:

Standard atmospheric pressure $= 1$ bar $= 14.5$ lbf/in^2

Note: 'Standard atmosphere' as defined by the physicist[2] is specified as a barometric pressure of 760 millimetres of mercury (mm Hg) at 0°C

1 standard atmosphere $= 1.01325$ bar $= 14.69$ lbf/in^2

The difference between these two standard pressures is a little over 1%. This can cause confusion. Throughout this book 1 bar is regarded as standard atmospheric pressure.

The torr is occasionally encountered in vacuum engineering.

1 torr $= 1$ mm Hg $= 133.32$ Pa

In measurements of air flow use is often made of water manometers.

1 mm of water (mm H$_2$O) $= 9.81$ Pa

References

1. BS 5514 *Reciprocating Internal Combustion Engines: Performance*
2. Kaye, G.W.C. and Laby, T.H. (1973) *Tables of Physical and Chemical Constants,* Longmans, London.

Further Reading

BS 350 Pt 1 *Conversion factors and tables*
BS 5555 *Specification for SI units and recommendations for the use of their multiples and of certain other Units*

1 The test cell as a system: an overall view

An engine test cell is a complex of machinery, instrumentation and services, all of which must work together as a whole. It is very easy for the specialist, concerned perhaps with a single topic such as combustion, engine control systems, exhaust emissions or noise, to fail to see the whole picture. In this chapter the authors put forward a methodology that they have found helpful as a means of pulling all the disparate elements together, and which is applied, explicitly or implicitly, throughout the book.

In the development of the theory of thermodynamics much use is made of the concept of the *open system*[1]. This is a powerful tool and can be very helpful in considering the total behaviour of a test cell. It is linked to the idea of the *control volume,* a space enclosing the system and surrounded by an imaginary surface, the control surface, Fig. 1.1. The great advantage of this concept is that once one has identified all the mass and energy flows into and out of the system it is not necessary to know exactly what is going on inside the system in order to draw up a 'balance sheet' of inflows and outflows.

The various inflows and outflows to and from a test cell are as follows:

In*	*Out*
Fuel	
Ventilation air	Ventilation air
(some may be used by the engine	Exhaust (includes air used by engine)
as combustion air)	Engine cooling water
Charge air (when separately supplied)	Dynamometer cooling water
Cooling water	Electricity from dynamometer
Electricity for services	Losses through walls and ceiling

Balance sheets may be drawn up for fuel, air, water and electricity, but by far the most important is the *energy balance,* since every one of these quantities has associated with it a certain quantity of energy. The same concept may be applied to the engine within the cell. This may be pictured as surrounded by its own control surface, through which the following flows take place:

* Compressed air may be a further energy input; however, usage is generally intermittent and it is unlikely to make a significant contribution

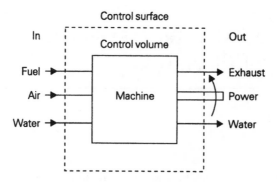

Figure 1.1. *An open thermodynamic system*

In	Out
Fuel	Power
Air used by the engine	Exhaust
Cooling water	Cooling water
Cooling air	Cooling air
	Convection and radiation

This is dealt with in Chapter 11, where the value of the method in the analysis of engine performance is made clear.

Table 1.1 shows a possible energy balance sheet for a cell in which a gasolene engine is developing a steady power output of 100 W. Note that where fluids (air, water, exhaust) are concerned the energy content is referred to an arbitrary zero, the choice of which is unimportant: we are only interested in the difference between the various energy flows into and out of the cell.

Of the heat in the exhaust gas leaving the engine, about 30 kW is transferred by radiation and convection from the exhaust system to the ventilation air in the cell.

The corresponding system is shown in Fig. 1.2, which also shows control surfaces for the engine and dynamometer.

If the power is absorbed by an electric dynamometer with regeneration the balance sheet looks slightly different, Table 1.2. Also in this case we have considered a diesel engine, with higher thermal efficiency. The corresponding system is shown in Fig. 1.3.

Table 1.3 shows the energy balance for a motoring test of a 100kW diesel engine. In this case there is of course no fuel consumption. The power developed by the dynamometer in driving the engine appears in the cooling water.

Table 1.1. *Test cell with hydraulic dynamometer and 100 kW gasolene engine*

Energy Balance

In		Out*	
Fuel	300 kW	Exhaust gas	60 kW
Ventilating fan power†	5 kW	Engine cooling water	90 kW
		Dynamometer cooling water	95 kW
		Ventilation air	70 kW
Electricity for cell services	25 kW	Heat loss, walls and ceiling	15 kW
	330 kW		330 kW

The energy balance for the engine, see Chapter 11, is as follows:

In		Out	
Fuel	300 kW	Power †	100 kW
		Exhaust gas	90 kW
		Engine cooling water	90 kW
		Convection and radiation	20 kW
	300 kW		300 kW

*This column shows the *increases* in energy of the air and cooling water in their passage through the cell taking an arbitrary zero datum at inlet conditions. For a detailed treatment see Chapter 11.

†The power to drive a forced draught ventilating fan appears ultimately as heat imparted to the ventilating air.

Notes
- The engine power output does not appear directly in the cell balance.
  —The majority appears as heat to the dynamometer cooling water.
  —The balance is represented by windage and heat convected from the dyna-mometer into the cell.
- A small proportion, perhaps 5%, of the ventilation air is drawn into the engine and appears as exhaust gas.
- The electricity for cell services, mostly lighting and fans, appears as heat in the ventilation air.
- The thermal efficiency of the engine

$$= \frac{\text{Power output}}{\text{Energy in fuel}} = 0.33$$

  a typical value for a gasolene engine, see Chapter 11.
- The *total* energy throughput is much greater than the power output of the engine.

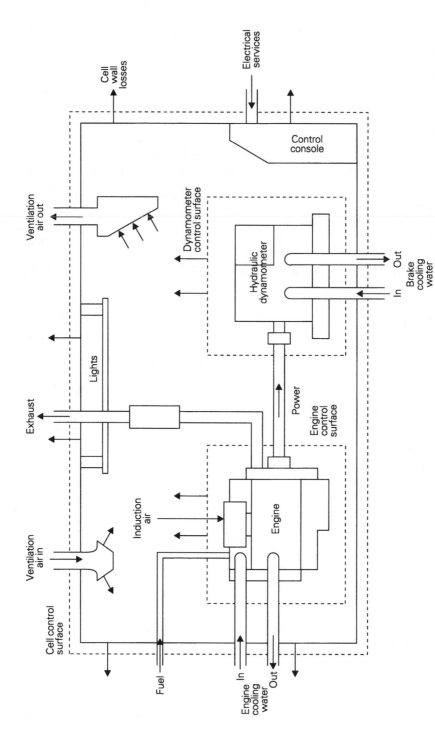

Figure 1.2. *Test cell with hydraulic dynamometer: an open system*

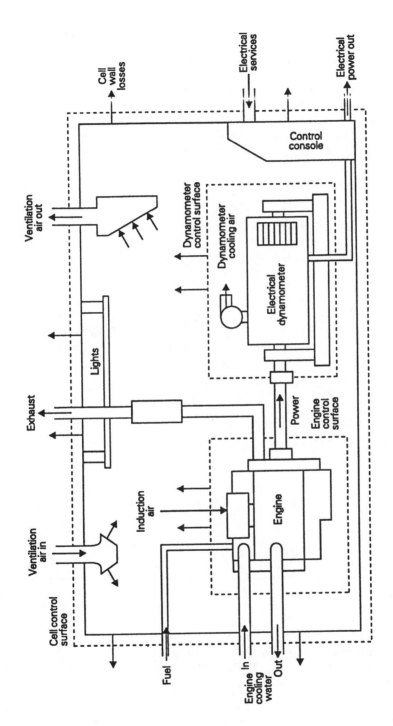

Figure 1.3. *Test cell with regenerative electrical dynamometer*

Table 1.2 *Test cell with regenerative electrical dynamometer and 100 kW diesel engine*

Energy balance			
In		*Out*	
Fuel	260 kW	Exhaust gas	45 kW
Ventilating fan power	5 kW	Engine cooling water	70 kW
Dynamometer excitation	5 kW	Dynamometer power	85 kW
		Ventilation air	80 kW
Electricity for cell services	25 kW	Heat loss, walls and ceiling	15 kW
	295 kW		295 kW

See Fig. 1.3 for the corresponding system.

In this case the energy balance for the engine is as follows:

In		*Out*	
Fuel	260 kW	Power	100 kW
		Exhaust gas	70 kW
		Engine cooling water	70 kW
		Convection and radiation	20 kW
	260 kW		260 kW

Notes
- Thermal efficiency of the engine = 0.385.
- Ventilation load is increased due to dynamometer air cooling load.

Table 1.3 *Motoring test of 100 kW diesel engine*

Energy balance			
In		*Out*	
Ventilating fan power	5 kW	Engine cooling water	25 kW
Dynamometer excitation	5 kW	Ventilating air	35 kW
Motoring power	30 kW	Heat loss, walls and ceiling	5 kW
Electricity for cell services	25 kW		
	65 kW		65 kW

Summary

The 'energy balance' approach outlined in this chapter will be found helpful in analysing the performance of an engine (Chapter 11) and in the design of test cell services (Chapters 4 and 6).

It is recommended that at an early stage in the design of a new test cell, diagrams such as Fig. 1.2 or Fig. 1.3 should be drawn up and labelled with flow and energy quantities appropriate to the capacity of the engines to be tested.

The large quantities of ventilation air, cooling water, electricity and heat that are involved will often come as a surprise. Early recognition can help to avoid expensive wasted design work by ensuring that:

- the general proportions of cell and services do not depart too far from accepted practice (any large departure is a warning sign);
- the cell is made large enough to cope with the energy flows involved;
- sufficient space is allowed for such features as water supply pipes and drains, air inlet grilles, collecting hoods and exhaust systems.

An example of this approach is given in Chapter 6.

Reference

1. Eastop, T.D. and McConkey, A. (1993) *Applied Thermodynamics for Engineering Technologists,* Longmans, London.

Further reading

Heywood, J.B. (1988), *Internal Combustion Engine Fundamentals,* McGraw Hill, Maidenhead.

2 Test cell design: an overall view

This chapter deals in broad outline with salient features of the main types of engine test cell, ranging from the simplest possible cell for the user with an occasional requirement to run a test, to complex multiple installations for the vehicle manufacturer and major oil company. Questions involved in the sizing of test cells, the provision of services, engine handling and safety and fire precautions are discussed.

Health and safety[1,2]

The engine test cell is an inherently dangerous working environment and this must be appreciated by everyone concerned with its design or operation. The walls, roof and floor of the cell must 'contain' the hazards of this environment. Openings, such as doors and ventilation ducts, must not compromise this function. Engine test cells are hot and noisy, with slippery floors and a working space full of pipes and cables. The sound attenuation between the control area and the cell will be enough to make calls for help difficult to hear.

The engine under test will not have been designed for installation in the cell, but for a quite different situation. It has probably not been designed with a view to safe containment of the effects of mechanical failure, such as may occur under test conditions. The shaft coupling the engine to the dynamometer and indeed the dynamometer itself may fail when testing a prime mover of novel design to its limits. The designer must foresee the possible results of such failures.

The presence of potentially explosive fuels in such an environment is clearly a hazard, and local fire regulations should be discussed with the relevant authorities at the design stage.

Before starting the design: some preliminary questions

Once construction has started design changes are an expensive matter. The following questions need careful consideration and a definitive answer right at the start.

1. What are the purposes for which the cell is intended?

What possible further purposes may be foreseen? Is confidentiality an important consideration? Independent consultants may be working for companies that are in competition and separation of control rooms will be necessary.

2. May there be a future requirement to install large additional equipment (e.g. an exhaust particulate tunnel) and how will this affect space requirements?
3. How will the engines be handled and mounted? How often will engines be changed and what arrangements be made for transport into and from the cells?
4. How many different fuels are likely to be in regular use?
   Must arrangements be made for special or reference fuels?
5. Are sufficient supplies of electrical power and water available on site? Water quality? Is regeneration of electric power (mains feedback) permitted?
6. What about the environmental impact? Will engine and exhaust noise present a problem?
7. Have all local regulations (fire, safety, environment, working practices, etc.) been studied?

Some typical test cell designs

The basic minimum

There are many situations in which there is a requirement to run engines under load, but so infrequently that there is no economic justification for building a permanent test cell. Examples are organizations concerned with engine over-haul and rebuilding, and specialist engine tuners.

To meet needs of this sort all that is required is a suitable area provided with:

- water supplies and drains
- portable fuel supply system
- adequate ventilation
- arrangements to take engine exhaust to exterior
- minimum necessary sound insulation
- adequate fire and safety precautions.

Figure 2.1 shows a typical installation, consisting of the following elements:

- portable test stand for engine and dynamometer
- engine cooling system
- engine exhaust system
- control console.

A 'bolt-on' dynamometer, see Chapter 7, requires no independent founda-tion, and is hence particularly suitable for a simple installation. This type of

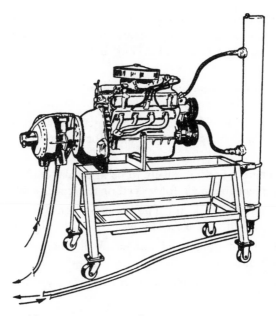

Figure 2.1.   *Portable engine test stand*

dynamometer is bolted to the engine bell housing, using an adaptor plate, with a splined shaft connection engaging the clutch plate. Occasionally the dynamometer may be installed without removal of the engine from the chassis.

The engine cooling system best includes a cooling column, p. 103, which minimizes the quantity of cooling water run to waste. This is much preferable to running cold mains water through the engine jacket. In some cases the radiator associated with the engine under test may be used.

The control console requires, as a minimum, indicators for dynamometer torque and speed, dynamometer flow control valve, engine throttle and stop controls, oil pressure gauge and, preferably, a simple fuel consumption gauge. A manually operated unit of the 'burette' type is adequate.

Typical applications

- proving of engines after overhaul or rebuild
- tuning of engines for racing etc.
- military vehicle overhaul
- checking emissions against legislative requirements.

General purpose engine test cells, 50 to 300 kW

This category represents by far the largest number of test cells in service. See

Fig. 2.2 for a typical layout and Fig. 2.3 for a dynamometer and engine mounting stand.

Such cells are often built side-by-side in a line. Engines enter the cell by way of a large door in the rear wall while the operator may enter by way of a door in the front wall to one side of the control desk. There is a double-skinned toughened glass window in front of the control desk. Most wall-mounted instrumentation, smoke meters, fuel consumption meters, etc., is carried on the side wall remote from the cell access door.

In cells rated at above about 150 kW it is usual to provide a crane rail located above the test bed axis with a hoist of sufficient capacity to handle engines and dynamometer. Signal conditioning units for the various engine transducers (pressure, temperature, etc.) are usually housed in a box carried by an adjustable boom. Multi-core cables communicate with the data acquisition unit in the control desk.

Boom boxes can be of considerable size, with contents that increase in the process of cell development, and their weight must be controlled to prevent overload at the wall fixing point.

In the cell illustrated, (Fig. 2.2) flexible rigging pipes lead from the engine manifold to the permanent exhaust and silencing system. Each exhaust outlet should include a butterfly valve for control of exhaust back pressure. The

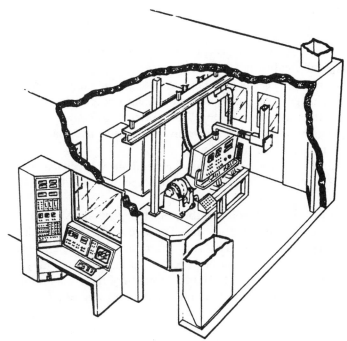

Figure 2.2. *General purpose engine test cell*

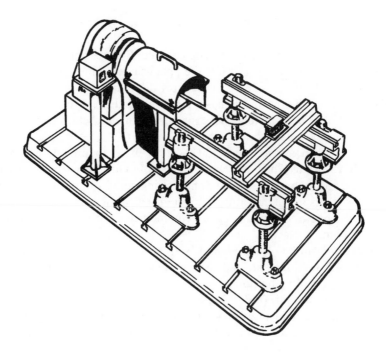

Figure 2.3.   *Dynamometer and engine mounting stand*

design of a test cell exhaust system can be a complex problem, see Chapter 6, p.112, for a full treatment.

The cell should be easily adaptable to take a wide range of engines, but the number of changes is typically not more than one per week. An adjustable engine mounting as shown in Fig. 2.3 is often provided as an alternative to individual engine mountings; such universal stands give maximum flexibility but set-up times may be rather long.

The control system associated with a general purpose cell may vary widely in the degree of complexity, depending on the type of testing for which the cell is intended, see Chapter 3.

Details of course vary considerably, but this general layout may be seen in thousands of installations world-wide. The layout of Fig. 2.2, with the test bed axis coinciding with the cell axis, is the more usual one. The alternative, with the test bed axis lying transversely, Fig. 2.4, gives better visibility, but special attention must be paid to safety and the strength of the observation window. It is sometimes convenient, with multiple installations, for adjacent cells to share a common control desk.

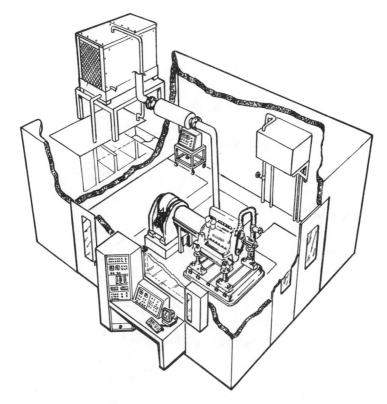

Figure 2.4.   *General purpose engine test cell: an alternative arrangement*

Typical applications

- development departments of engine and component manufacturers
- independent testing and development laboratories
- fuel and lubricant development and standard test procedures
- training and educational facilities
- military workshops.

Specialized test cells for research and development

Engine test and development facilities represent a major element of investment for vehicle manufacturers and large oil companies. There will in general be a number of similar cells at any site, standardized as far as possible, and to achieve a high load factor engines are usually mounted on standard pallets and pre-rigged before transfer to the test cell, where quick release couplings ensure a rapid turn-round, Fig. 2.5.

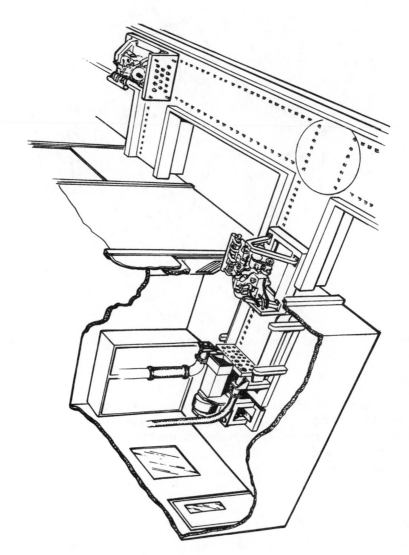

Figure 2.5. *Test cell with pallet mounted engine.*

Multiple installations of this kind call for elaborate ventilation, cooling water and exhaust collection systems.

Typical applications

- development departments of vehicle manufacturers and major oil companies
- government testing and monitoring laboratories.

Inclined test beds

With the increased use of 'off the road' vehicles there is likely to be an occasional requirement for inclinable test beds capable of handling engines running with the crankshaft centreline inclined to the horizontal. This may present problems with hydraulic dynamometers with open water outlet connections. Closed circuit systems, as are usual with eddy current machines, are more easily adapted to inclined running.

For the very occasional case of the engine with vertical crankshaft e.g. outboard boat engines, the electrical dynamometer is the obvious choice and may generally be used without modification. Special arrangements need to be made for torque calibration.

Production Test Cells

These cells are highly specialized installations lying rather outside the scope of this book. The objective is to check that the engine is complete and runnable in the minimum possible time: typical floor-to-floor times for a gasolene engine are 6.5 min and for a diesel engine for a goods vehicle 20 min.

The whole procedure – engine handling, rigging, filling and draining and the actual test sequence – is highly automated, with interventions by the operator limited to dealing with faults and perhaps to checking for leaks and abnormal noise. The test cell is programmed to recognize variants of a standard engine and to adjust the pass or fail criteria accordingly.

Typical measurements made during a production test include:

- time taken for engine to start
- cranking torque
- time taken for oil pressure to reach normal level
- exhaust gas composition.

Most gasolene engines are no longer loaded during a 'hot test'. This is not the case with diesel engines, for which power output is measured and recorded.

Cold testing is sometimes applied to partially built engines. The engine is driven electrically and measurements made of:

- cranking torque
- oil pressure
- compression pressure.

Checks are made for leakage, but this cannot of course cover leaks at operating temperature. The justification for such tests is mainly reduced cost, as neither a fuel supply nor an exhaust system is required.

In the design phase of a production test facility a number of fundamental decisions have to be made, including:

- layout, e.g. conveyor loop with work-stations, carousel
- what remedial work, if any, to be carried out on test stand
- processing of engines requiring minor/major rectification
- engine handling system, e.g. bench height, conveyor and pallets, 'J' hook conveyor, automated guided vehicle
- engines rigged and stripped at test stand or remotely
- maintenance facilities and system fault detection
- measurements to be made, handling and storage of data.

Production testing imposes heavy wear and tear, particularly on engine rigging components, and these can become a maintenance liability. Modular construction of key sub-assemblies will allow repairs to be carried out by replacement of complete units.

Overall size of cell

A cramped cell in which there is not room to move around in comfort is a permanent source of danger and inconvenience. The smaller the volume of the cell the more difficult it is to control the ventilation system under conditions of varying load, see Chapter 4. As a rule of thumb there should be an unobstructed walkway 1 m wide all round the test bed while the cell height should be sufficient to allow the dynamometer to be lifted over the top of the engine. It is often necessary, when testing vehicle engines, to accommodate the exhaust system as used on the vehicle, and this may call for extra length in the cell.

It must be remembered that much of the plant in the cell requires calibration from time to time and there must be adequate access for the calibration engineer and his instruments. The major problem may be the calibration of the dynamometer, involving accommodation of a torque arm and dead weights.

Seeing and hearing the engine

Except in the case of production test beds it is almost universal practice to separate the control room from the cell proper. This means that the operator at the desk has only a restricted view of the engine, but this is a less serious matter than it once was, thanks to modern instrumentation and closed circuit television. Comprehensive alarm systems make observation less critical and simple aids such as wide-field mirrors can be of use.

The importance or otherwise of visibility is linked with a fundamental question: which way round is the cell to be arranged? It is obviously more convenient from the point of view of engine installation to have the engine at the rear, adjacent to the access door, but this gives the worst visibility.

The experienced operator will be concentrating attention on the indications of instruments and display screen, and will only catch changes happening in the cell out of the corner of an eye. It is well to avoid possible distractions, such as dangling labels or identity tags that can flutter in the ventilation wind.

Hearing has always been important to the experienced test engineer, who can often detect an incipient failure by ear well before it manifests itself in any other way. Unfortunately modern test cells, with their generally excellent sound insulation, cut off this source of information and consideration should be given to the provision of in-cell microphones with external loudspeakers or earphones.

Flooring and sub-floor construction

Fire regulations invariably call for spaces below floor level to be scavenged by the ventilating system to avoid any possibility of the build-up of explosive vapours. Sometimes there is a statutory requirement for the air quality to be monitored. Fuel services should preferably not be run in floor trenches. It is good practice to provide floor channels on each side of the bed, as they are particularly useful for running 'cold' services and drains in an uncluttered manner.

Floor channels should be covered with well-fitting chequer plates, not weighing more than about 20 kg each, and provided with lifting holes. The plates can be cut as necessary to accommodate service connections.

The floor, or seismic block when fitted, must be provided with arrangements for bolting down the engine and dynamometer. The best solution is to cast in position in the concrete two or more cast-iron T-slotted rails. The machined surfaces of these rails form the datum for all subsequent alignments and they must be set and levelled with great care. Any twist could lead to serious distortion. The use of fabricated box beams is a false economy.

Sometimes complete cast-iron floor slabs with multiple T-slots are used but these tend to trap liquids and are highly sound-reflective. Also they can be very slippery. Floor finishes should not be so smooth as to be dangerous and particular care is necessary where fork-lifts or other transport are used. The same applies to any kind of sloping access.

Doors

Doors that meet the requirements of noise attenuation and fire containment are inevitably heavy and require more than normal effort to move them; this is a safety consideration to be kept in mind when designing the cell. Forced or induced ventilation fans can give rise to pressure differences across doors, possibly making it dangerous or impossible to open a large door.

All test cell doors must be either on slides or be outward opening. There are designs of sliding door which are suspended on rails and drop to seal in the closed position but all sliding doors have the disadvantage of creating 'dead' wall space when open. Doors should be provided with small observation windows and may be subject to regulations regarding the provision of EXIT signs.

Walls

Test cell walls are requred to meet certain special demands in addition to those normally associated with an industrial building. They, or the frame within which they are built, must support the load imposed by any crane installed in the cell, plus the weight of any equipment mounted on or suspended below the roof. They must be of sufficient strength and suitable construction to support wall mounted instrumentation cabinets, fuel systems, and any equipment carried on booms cantilevered out from the walls. They should provide the necessary degree of sound attenuation and must comply with requirements regarding fire retention (usually a minimum of one hour's containment).

High density building blocks[10] provide good sound insulation but usually require some form of internal acoustic treatment, such as 50 mm thick sound absorbent panels, to reduce the level of reverberation in the cell. Such panels can be effective on walls and ceilings, even if some areas are left uncovered for the mounting of equipment.

The choice of cell wall material has been complicated in recent years by the availability of special construction panels made of sound absorbent material sandwiched between metal sheets, of which the inner (cell) side is usually perforated. These 100 mm thick panels may be used with standardized structural steel frames for the construction of test cells and, while not offering the same level of sound attenuation as dense block walls, give a quick and clean method of construction with a pleasing finished appearance; this is particularly useful when cells are to be built in an existing building. However proper planning is necessary if heavy equipment is to be carried, as internal 'hard-mounting' points must be built into the steel structure.

Lighting[3]

The typical test cell ceiling is cluttered with fire sprinkler systems, exhaust outlets, ventilation ducting and a lifting beam. The lights are often the last consideration, but are of vital importance. They must be securely mounted so as not to move in the ventilation 'wind' and give a high and even level of lighting without causing glare from the control room window. Since the lights may be working in an atmosphere of oil-laden fumes they must meet appropriate safety standards, be easily cleaned and operate at a moderate surface temperature.

The detailed design of a lighting system is a matter for the specialist. The 'lumen' method of lighting design gives the average level of illumination on a working plane for a particular number of 'luminaires' (light sources) of specified power arranged in a symmetrical pattern. Factors such as the proportions of the room and the reflectivity of walls and ceiling are taken into account.

The unit of illumination in the International System of Units is known as the *lux*, in turn defined as a radiant power of one *lumen* per square metre. The unit of luminous intensity of a (point) source is the *candela*, defined as a source which emits one *lumen* per unit solid angle or 'steradian'. The efficiency of light sources in terms of candela per watt varies widely, depending on the type of source and the spectrum of light that it emits.

The IES Code[9] lays down recommended levels of illumination in lux for different visual tasks. A level of 500 lux in a horizontal plane 500 mm above the cell floor should be satisfactory, but areas of deep shadow must be avoided. Emergency lighting with a battery life of at least one hour and an illumination level in the range 30–80 lux should be provided in both test cell and control room.

It is sometimes very useful to be able to turn off the cell lights from the control desk, whether to watch for sparks or red-hot surfaces or simply to hide the interior of the cell from unauthorized eyes. An outlet for a 'wandering' inspection lamp should be included in the layout. For reasons of safety the supply to such lamps should not exceed 110 V.

Engine handling systems

The degree of elaboration of the system adopted for installing and removing the engine naturally depends on the frequency with which the engine is changed. At one extreme, a large marine engine will be built in the position in which it will finally be tested. In some research cells the engine is more or less a permanent fixture but at the other extreme the test duration for each engine in a production cell may be measured in minutes and the time taken to change engines must be cut to an absolute minimum.

There is a corresponding variation in the handling systems:

• Simple arrangements when engine changes are comparatively infrequent.

The engine is mounted on a suitable stand which is then lifted into the cell. All connections are made subsequently.

- The engine is mounted on a wheeled trolley carrying various transducers and service connections which are coupled to the engine before it is moved into the cell.
- For production test beds it is usual to make all engine connections prior to entering the cell. The pallet carrying the engine is wheeled into the cell where an automatic or semi-automatic docking system permits all connections (including in some cases the driving shaft) to be made in seconds. Such systems are particularly useful when the test beds are required to deal with a number of different engine types, possibly presented in random order.

Engine rigging

A considerable number of connections must be made to any engine under test: the coupling shaft, fuel, cooling water, exhaust and a wide range of transducers and instrumentation. Workshop support in the provision of suitable fittings and adaptors should always be made available. All pipe connections should be flexible and exhaust connections can be particularly troublesome and short-lived at high temperatures; they should be treated as consumable items. Transducers are difficult to fit in some kinds of flexible tubing.

Transducer leads must be marshalled neatly and kept away from hot surfaces. Boom-mounted instrumentation boxes may be overheated if mounted above the engine and must be positioned suitably.

Shaft guards

The guarding of shafts and couplings presents the cell designer with a dilemma. A guard sufficiently strong to withstand a catastrophic failure is likely to be so massive and inconvenient to fit that it will fall out of use. In practice the guard merely prevents personnel from coming into contact with the rotating parts.

Occasionally guards are made with a close-fitting wooden liner to suppress possible shaft whirl. Such guards must be carefully positioned and checked for possible rubbing under starting, running and idle conditions.

In-cell controls

There are a number of tasks that require an operator to be in the cell while the engine is running. Examples are the tuning of racing car engines, the setting of idling adjustments and checking for leaks. It is very desirable, for safety reasons, that there should always be a second operator at the control desk.

A simple control station within the cell, enabling the operator there to control the engine and observe torque and speed, is often useful but must be interlocked to prevent control actions being taken outside the cell at the same time. Duplication of all control desk displays within the cell is seldom justified.

Emergency stop and engine shut down

The engine test cell control system must include a robust method of shutting down the engine.

All test facilities must be provided with a number of emergency stop buttons, of large size and conspicuously marked, both at the control desk and at strategic positions in the cell. An emergency stop button is intended for use in the case of a local or single cell emergency other than a fire.

The traditional function of the emergency stop circuit is to shut off electrical power from all devices directly associated with the running of the engine. Dynamometers are usually switched to the safe or 'de-energized' condition, fuel supply isolation systems are activated, ventilation fans are shut down, but services shared with other cells and fire dampers are not interrupted.

Since spark ignition engines may be stopped by interrupting the supply to the ignition circuit this is usually linked to the emergency stop system. A diesel engine calls for a more robust shut down system: while many automotive diesels are fitted with a shut down solenoid, which should be wired into the emergency stop system, it should be backed up by some means of independently moving the fuel pump rack to the stop position. Some test plant suppliers produce standard devices or modified throttle controllers which perform this function, using stored pneumatic power as a precaution against electrical failure.

There is always the possibility of a diesel engine running away, usually as a result of being fuelled by lubricating oil: this puts it out of the control of the fuel pump. Such an occurrence is rare but not unknown, even with modern engines. Horizontal rail or bus engines, if inadvertently overfilled with lubricating oil, are prone to this trouble. If the emergency stop system is activated when an engine is running away all load is removed from electrical or eddy current dynamometers and the situation is made worse.

The two ways of stopping the engine in this situation are either by closing off the combustion air supply or applying sufficient load to stall the engine. Unless the engine has been specially rigged with a closure valve or a means of injecting CO_2 the former method is often not practicable. Stalling the engine in such an emergency calls for a cool head and is best arranged by providing a special 'fast stop' button which ramps the dynamometer to full torque and de-energizes the engine systems. Such rapid stalling of the engine can be damaging: a fast stop button should only be located at the control desk and operators should be carefully briefed on its use.

Fire control

The fire alarm and fire extinguishing systems are always entirely independent of the emergency stop arrangements and are linked to the ventilation control system since in the event of a fire in the cell it is clearly undesirable to continue to force in ventilating air. It is also undesirable to dowse the valuable contents of the cell without good cause.

There is much legislation relevant to industrial fire precautions[1,2] and also a number of British Standards[4-8]. The Health and Safety Executive, acting through the Factory Inspectorate, are responsible for regulating such matters as fuel storage arrangements, and should be consulted, as should the local Fire Authority.

Fire extinguishing systems

The following information on fire extinguishing arrangements has been provided by the School of Fire Safety Engineering of the Fire Service College, Moreton-in-Marsh, U.K.

Carbon dioxide (CO_2)

CO_2 can be used against flammable liquid fires. CO_2 is about 1.5 times heavier than air and it will tend to settle at ground level in enclosed spaces. The discharge of CO_2 is likely to be very noisy and misting of the atmosphere will take place. It should be noted that if CO_2 is used in confined spaces or in large quantities, as in a total flood system, concentrations may reach levels that endanger life. Breathing difficulties become apparent above a concentration of 4% and a concentration of 10% can lead to unconciousness or, after prolonged exposure, to death.

Dry powder

Powders are designed for high-speed extinguishment of highly flammable liquids such as petrol, oils, paints and alcohol; they can also be used on electrical or engine fires. It must be remembered that dry chemical powder does not cool nor does it have a lasting smothering effect and therefore care must be taken against re-ignition.

Halons

Following the Montreal Protocol, some halons (halogenated hydrocarbons) are being phased out. These contain chlorine or bromine, thought to be damaging to the ozone layer, and their production has been banned. Unfortunately this ban embraces Halon 1211 and Halon 1301, both hitherto used for total flooding and in portable extinguishers for dealing with flammable liquid fires. It is still feasible to use existing stocks, but clearly there are strong arguments for seeking replacements. Some are available, mostly fluorinated hydrocarbons, but care must be exercised as they may have toxic effects at their design concentrations

Inergen

Inergen^{TM} is the trade name for the extinguishing gas mixture of composition 52% nitrogen, 40% argon, 8% carbon dioxide. There are other alternative extinguishants, including pure argon, many of which may be used in automatic mode even when the compartment is occupied, provided the oxygen concentration does not fall below 10% and the space can be quickly evacuated.

With all gaseous systems, precautions should be taken to ensure that accidental or malicious activation is not possible. In carbon dioxide systems, automatic mode should only be used when the space is unoccupied.

Total flood systems, of whatever kind, are usable only after the area has been evacuated and sealed. They must be interlocked with the doors and special warning signs must be provided.

Foam

Foam extinguishers could be used on engine fires but they are more suited to flammable liquid spill fires or fires in containers of flammable liquid. If foam is applied to the surface or sub-surface of a flammable liquid it will form a protective layer. Some powders can also be used to provide rapid knock-down of the flame. Care must be taken, however, since some powders and foams are incompatible.

Micro-fog systems

Micro-fog, like other water-based fire extinguishing systems, has the great advantage that it removes heat from the fire source and its surroundings and thus reduces the risk of re-ignition when it is switched off. Micro-fog systems use very small quantities of water (typically 5 litres) and discharge it as a very fine spray. They are particularly efficient in large cells, such as vehicle anechoic chambers, where the fire source is likely to be of small dimensios relative to the size of the cell.

Vapour detection

Devices are available for both oil mist and hydrocarbon vapour detection. However their use presents a practical problem in setting the sensitivity to a level that warns of real danger without continuously tripping at a level which experience shows to be safe.

Summary

The intrinsically dangerous nature of the test cell environment is emphasized and emergency stop and fire precautions are described. Before starting to design a cell important decisions must be taken. Incorrect decisions at this stage can be very expensive.

There are a number of different basic types of test cell, depending on the use for which they are intended. Over- and under-provision of facilities should both be avoided.

The special features of test cell construction, which differs from that of any other kind of industrial building, are described.

References

1. Ridley, J. (1998) *Safety at Work,* 4th edn, Butterworth-Heinemann, Oxford.
2. Arscott, P. (1980) *An Employer's Guide to Health and Safety Management,* Engineering Employer's Federation.
3. BS 8206 Part I *Code of Practice for Artificial Lighting.*
4. BS 5306 Parts 0 to 7 *Fire Extinguishing Installations and Equipment on Premises.*
5. BS EN3 Parts 1 to 6 *Specification for Portable Fire Extinguishers.*
6. BS 5997 *Guide to British Standard Codes of Practice for Building Services.*
7. BS 6959 *Code of Practice for Selection, Installation, Use and Maintenance of Apparatus for the Detection and Measurement of Combustible Gases.*
8. BS EN 137 *Specification for Respiratory Protective Devices: Self-contained Open-circuit Compressed Air Breathing Apparatus.*
9. The IES Code, Chartered Institution of Building Services, London.
10. BS 6073 Parts 1 and 2 *Precast Concrete Masonary Units.*

3 Controls and the control room – the electrical installation

This chapter deals with the other main component of an engine testing installation: the control room. Recommendations are made regarding the design of control rooms and an analysis is made of various methods of achieving control and problems that may be encountered. Attention is drawn to the detailed regulations that govern every aspect of industrial electrical supply systems. No attempt is made to summarize them, but aspects peculiar to the test cell and control room environment are considered. Of these perhaps the most important concerns electrical interference between power and signal circuits[1]. Finally, the process of drawing up a complete test cell and control room specification is discussed.

Figure 3.1 shows a typical control room layout. As the operator will spend a great deal of time in the control room its layout should be as convenient and comfortable as possible. There should be a working surface away from the control station, where notes may be written and curves plotted and discussions may be held. A keyboard will be needed, but not all the time, and it should be kept in a drawer when not in use. Some equipment often installed in control rooms, such as emissions instrumentation, can have a quite large power consumption. This must be taken into account when specifying the ventilation system, which should conform to office standards.

The control console itself calls for careful design on sound ergonomic principles, and must initially include space for later additions. Figure 3.2 shows a typical general purpose control desk[2].

By definition, the primary purpose of the control desk is either to permit the operator to control the engine directly, or to allow him to manage and monitor a control system that may be automated to a greater or less degree. Pride of place will thus be given to the controls that govern engine torque and speed, the prime variables, and to the corresponding indicators of these variables, which may take various forms. Of almost equal importance are secondary indications and warning signals covering such features as coolant temperature, lubricating oil pressure, exhaust temperature, etc.

The size and position of the observation window greatly affects the control room layout. In special climatic or anechoic cells windows are not fitted because they compromise the integrity of the structure. Modern CCTV equipment can make a window into the cell unnecessary and greatly simplify control area layout, often dominated by the problem of finding window space.

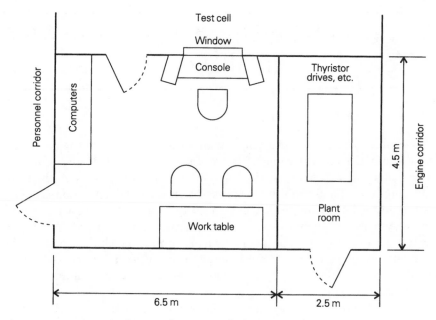

Figure 3.1.   *A typical control room and plant room layout*

The window should not be located opposite the cardan shaft assembly, the most likely source of flying debris, and will consist of two sheets of toughened glass, typically about 10 cm apart[3]. It should be large enough to give good visibility over most of the cell area but not so large that instrument cabinets on each side are at an inconvenient distance from the operator. The larger the window the greater the problem of achieving sound attenuation between the cell and control room. The instrument cabinets are commonly 19 in rack units up to 6 ft high and should be set with their faces at right angles to the operator's line of sight and with frequently used controls at a convenient height.

Displays and controls for secondary equipment such as cell services can be relegated to lower levels in the cabinets, while instruments used only inter-mittently, such as fuel or smoke meters, should be installed towards the top and operated if necessary from a standing position. The positioning of graphic displays or computer monitors can be a problem, as they occupy considerable space and the solution may be to use articulated support booms.

The choice of indicator depends on the type of control and the kind of work to be done. If, as is still often the case, the operator needs to manage the engine manually by a T-handle or rotary dial type throttle control he should have an analogue speed indicator prominently in view.

In general the analogue indicator, in which a pointer moves over a

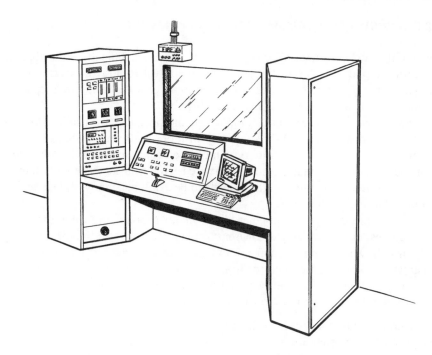

Figure 3.2. *General purpose control desk*

graduated dial, is the appropriate choice when the quantity indicated is under manual control.

If the indicator is required to show the stage reached in a test, current levels of such quantities as power and speed and any departures from desired values, a digital readout, perhaps accompanied by warning signals, is appropriate.

Presentation of readings in the form of a row of coloured bars on a screen can be useful if a number of inter-related quantities are to be kept under review.

Finally a continuous recording, either on the computer screen or as a pen recording, may be the best solution. For some tests, particularly of transient phenomena, it may be useful to be able to 'freeze' such recordings. For example, if turbocharger lag during acceleration were being investigated it would be useful to see a simultaneous plot of throttle position, turbocharger speed, boost pressure, exhaust pressure and engine torque over a period of perhaps 10 s.

The operator's desk should be designed so that liquids spilled on the working surface cannot run down into electronic equipment mounted beneath and control rooms should have two emergency escape routes. The floors should have non-slip and if possible anti-static surfaces.

Instruments and controls: planning the layout

Even if the above principles are kept firmly in mind it is difficult, in the absence of a formal plan, to prevent the test cell and control room from degenerating into a jumble of different devices positioned more or less at random as space is available. This is not only inefficient but unsafe[4].

Computerization has greatly complicated the problem. In 'traditional' test cells it was unusual to see a direct display of more than perhaps ten or twelve quantities with possibly ten more on a multi-pen chart recorder. In computer-controlled cells, particularly those associated with transient testing, Chapter 15, it is not unusual for up to 128 channels to be available, many of them giving rapidly changing indications.

This can present the cell designer with a major problem in seeing 'the wood for the trees' and organizing a coherent display. Raw data in the form of long columns showing channel numbers and associated current values is not the ideal way of keeping the operator in touch with what is going on.

It may be found helpful, in planning the layout of instruments and controls, to classify them under the following headings:

- operational and safety instrumentation
- primary instrumentation
- secondary instrumentation.

If the status or output of most or all of the instrumentation is shown on one or more Visual Display Units (VDU's) these displays must be arranged so that those of high priority are visually prominent.

Operational and safety instrumentation

This term covers the basic instruments and controls necessary for the safe running of the engine, whatever the purpose of the test. Operational instrumentation is commonly 'hard wired'*, with duplicate signals going to any associated computer, and includes devices associated with the cell shutdown procedure. All the following indications would normally be included under this heading:

- engine speed
- oil pressure
- cooling water flow and temperature
- signals associated with fire suppression and detection

---

* The term 'hard wiring', as used in this book, means that the instrument or other device receives its signal via its own unique cable rather than by way of a computer (software-dependent) signal or a multiplexing system.

- in-cell beacon/alarm automatically triggered by starting sequence
- status of cell ventilation, purge system and combustion air supply
- overspeed trips on *both engine and dynamometer* to deal with cardan shaft failures

Primary instrumentation

This covers the essential data relating to the experimental purpose of the cell. The display of this information will be arranged in a prominent position where it is comfortably seen from the operator's desk.

The definition of primary as opposed to secondary instrumentation depends entirely on the purpose of the cell. In a cell intended for noise measurement studies, for example, sound level indicators and recorders have pride of place while indications of engine torque, exhaust temperature or fuel consumption are of secondary importance.

A test cell devoted to diesel engine exhaust emissions has as primary instrumentation the required gas analysis system, with all the associated controls of the gas dilution and sampling system. If, as is often the case, the engine is being put through a computer-controlled transient drive cycle, information regarding engine state is of secondary importance and displayed accordingly.

Secondary instrumentation

Information not of direct relevance to the test should be displayed away from the operator's normal field of attention, or it may be recorded on selectable pages of the main VDU display. Typical secondary information concerns slowly changing quantities such as barometric pressure, relative humidity and cooling water temperature.

Exhaust emission measuring equipment

The housing of emission measurement equipment presents a major problem in the layout of the facility; this is the case whatever the type of engine for which the cell is intended. Equipment has tended to become smaller in recent years with the advent of 'mini-tunnels' for the measurement of particulates but there are still large cabinets to be housed and restrictions on the length of sampling tubes from exhaust pipe to tunnel.

There are essentially four types of equipment concerned, each with particular requirements regarding accommodation:

- equipment housed in test cell and connected to the exhaust system by way of heated pipeline

- equipment housed in control room or in an annex. There are usually restrictions on the maximum distance between this apparatus and the in-cell equipment. Cabinets may be mounted on castors, giving some flexibility, but may need to be connected to a vent system to atmosphere
- a bottle store for the reference gases with appropriate manifold and control valves. Gas bottles are heavy and the store is usually at ground floor level in a secure area with good access for bottle changing. It is of course possible for several cells to use a common bottle store but some gases require high grade stainless steel piping and the layout of any distribution system needs careful consideration to minimize costs. The bottle store area may require forced ventilation.
- post processing equipment including laboratory scales for weighing filter papers. There are strict rules covering the handling of the filter papers used in particulate measurement and diesel engine homologation. Some facilities set aside a special laboratory while others house the samples and scales in climatically controlled cabinets in the control room.

Alarm monitoring

This can be a complex matter and is covered in more detail in Chapter 17.

Primary alarms should refer to limits which, if exceeded, would endanger the equipment; these are not confined to the engine but will include the support systems and may have their status shown on a dedicated alarm display panel in addition to being part of a computer screen display.

Included in this classification are the hard-wired safety alarms relating to such matters as fire, overspeed, water supply, engine oil pressure and the emergency stop system.

Secondary alarms not of importance in a particular case may be set to wide limits or even switched off to avoid overloading the operator. It is desirable that alarm signals should be logged in an 'historic log' buffer.

Security of supply

The possible consequences of a failure of the electricity supply, or of such factors as undetected fuse failures, must be carefully thought out. It is usual to design safety systems as normally energized (de-energize to trip), but all aspects of cell operation must be considered. For instance solenoid-operated valves must latch in the correct position when power fails.

Particular dangers can arise if the power is restored unexpectedly after a shut-down. For example a power failure can result in the throttle actuator moving either to maximum or minimum, depending on the linkage.

System status displays must indicate the actual rather than the requested status (the cause of the Three Mile Island disaster).

The installation of an uninterrupted power supply (UPS) for critical control and data acquisition devices is to be recommended. This relatively inexpensive piece of equipment need only provide power for the few minutes required to save or transfer computer data.

Test cell start-up and shut-down procedures

It has already been remarked that an engine test cell is a potentially dangerous place. Failure of any test cell service, or failure to bring the services on stream in the correct order, can be disastrous, and it is essential to devise a foolproof system for start-up and shut-down, with all necessary alarms and automatic shut-downs. This system should preferably be hard wired and should be quite separate from the control system for engine and dynamometer, while interacting with it at all necessary points.

A visual inspection *must* be made before each and every test run. Long-established test procedures can give rise to complacency and lead to accidents.

Checks before start-up

It will be assumed that routine calibration procedures have been followed, and that instrument calibrations are correct and valid.

- No 'work in progress' or maintenance labels are attached to any system switches.
- Engine/dynamometer alignment within set limits and shaft bolts tightened to correct torque.
- Shaft-guard in place and centred so that no contact with shaft is possible (if appropriate, it is a good practice to rock the engine on its mounts to see that the rigging system, including exhaust tubing, is secure and flexes correctly).
- All loose tools, bolts, etc. removed from the test bed.
- Engine support system tightened down.
- Fuel system connected and leak proof.
- Engine oil at the correct level and oil pressure alarm hard wired into control system.
- All fluid services such as dynamometer water are on.
- Fire system primed.
- Ventilation system available and switched on to purge cell of flammable vapour prior to start up of engine.

- Check that cell access doors remote from the control desk have warnings set to deter casual entry during test.

Checks immediately after start-up

- Oil pressure over-ride to be released when pressure is above trip setting.
- If stable idle is possible and access around the engine is safe a quick in-cell inspection should be made. Particularly look for engine fuel or oil leaks, cables or pipes chafing or being blown against the exhaust system and listen for abnormal noises.
- *Test the emergency shut-down system by operating any one of the stations.*
- Re-start and run test.

Checks immediately after shut-down

- Allow cooling period with services and ventilation left on.
- Shut off fuel system as dictated by site regulations (draining down of day-tanks etc.).
- Carry out data saving, transmission or back-up procedures.

The test sequence: modes of control[2]

It should be recognized that a test programme for an engine coupled to a dynamometer is first and foremost a sequence of desired values of engine torque and speed. This sequence is achieved by manipulating only two controls: the engine 'throttle' and the dynamometer torque setting. (It is convenient to refer to the throttle, even when, as in the case of a diesel engine, this control in fact acts, usually by way of some form of governor, on the fuel pump rack setting or equivalent electronic control.) The dynamometer torque is set by the position of the sluice plates or by control of one or both inlet and outlet valves in the case of hydraulic machines, by a control on the excitation of eddy current machines or of the wave form generated by the thyristor drive of a.c. or d.c. machines.

For any given setting of the throttle the engine has its own inherent torque-speed characteristic and similarly each dynamometer has its own torque-speed curve for a given control setting. The interaction of these two characteristics determines the inherent stability of the engine dynamometer combination.

The engine or throttle control may be manipulated in three different ways:

- to maintain a constant throttle opening   (position mode)
- to maintain a constant speed   (speed mode)

- to maintain a constant torque   (torque mode).

The dynamometer control may be manipulated:

- to maintain a constant control setting   (position mode)
- to maintain a constant speed   (speed mode)
- to maintain a constant torque   (torque mode)
- to reproduce a particular torque-speed characteristic (power law mode).

Various combined modes, usually described in terms of engine mode/dyna-mometer mode, are possible.

Position mode

This describes the classical 'engine test'. The throttle is set to a desired position, the dynamometer control similarly set, and the system settles down, hopefully in a stable state. There is no feedback: this is an open loop system. Figure 3.3(a) shows a typical combination, in which the engine has a fairly flat torque-speed characteristic at fixed throttle opening, while the dynamometer torque rises rapidly with speed, typical of most but not all machines. The two character-istics meet at an angle approaching $90°$, and operation is quite stable.

Certain designs of variable fill dynamometer with simple outlet valve control, see Chapter 7, may become unstable at light loads, see Fig. 3.3(b), and this can lead to hunting, or to the engine running away. The two characteristics meet at an acute angle. A friction brake, Fig. 3.3(c), operating at a given control setting, develops a torque almost independent of speed and is clearly unsuitable for loading an engine in position mode.

Position and power law mode

This is a variation on pure position mode, in which the dynamometer controller is manipulated to give a torque speed characteristic of the form

$$\text{brake torque} = \text{constant} \times \text{speed}^n$$

When $n = 2$ this approximates to the torque characteristic of a marine propeller and the mode is thus useful when testing marine engines. It is also a safe mode, tending to prevent the engine from running away if the throttle is opened.

Position and speed mode

In this mode the throttle position continues to be set manually but the dynamometer is equipped with an automatic controller which adjusts the torque absorbed by the machine to maintain the speed constant whatever

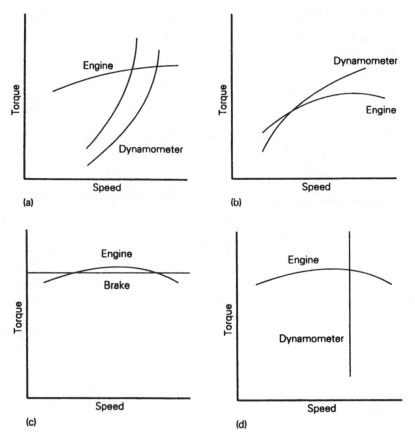

Figure 3.3. *Control modes, engines and dynamometers: (a) position mode, stable hydraulic dynamometer; (b) position mode, unstable hydraulic dynamometer; (c) position mode, friction brake; (d) position/speed mode; (e) position/torque, governed engine; (f) speed/torque mode; (g) torque/speed mode*

the throttle position and power output, Fig. 3.3(d). This is a very stable mode, and is generally used for plotting engine torque–speed curves at full and part throttle opening.

Position and torque mode (governed engines)

Governed engines (which includes most diesel engines) have a built in torque–speed characteristic, usually slightly 'drooping' (speed falls as torque increases). They are therefore not suited for coupling to a dynamometer in speed mode. They can however be run with a brake in torque mode. In this mode the

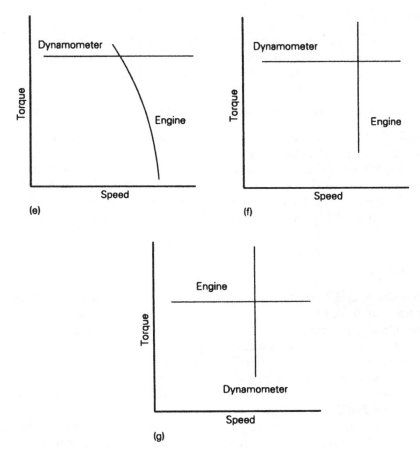

Figure 3.3.   *(cont.)*

automatic controller on the dynamometer adjusts the torque absorbed by the machine to a desired value, Fig. 3.3(e). Control is quite stable. Care must be taken not to set the dynamometer controller to a torque that may stall the engine.

Speed and torque mode

This is a useful mode for running in a new engine, when it is essential not to apply too much load. As the internal friction of the engine decreases it tends to develop more power and since the torque is held constant by the dynamometer the tendency is for the speed to increase. This is sensed by the engine speed controller, which acts to close the throttle, Fig. 3.3(f).

Torque and speed mode

This can be useful as an approximate simulation of the performance of a vehicle engine when climbing a hill. The dynamometer holds the speed constant while the engine controller progressively opens the throttle, Fig. 3.3(g).

Special features of the four-quadrant dynamometer

Where the dynamometer is also capable of generating torque it will be necessary to take precautions in applying some of the above modes of control. This applies when the dynamometer is in speed mode, Figs. 3.3(d) and (g). When running in either of these modes the four-quadrant dynamometer will seek to maintain the set speed even when the engine has ceased to deliver power. This could clearly be dangerous should the engine have failed in some way.

Throttle actuators

When the engine is under simple position (open loop) control the manual control may be directly linked to the engine throttle lever. Alternatively, and more usually, the engine lever is moved by a throttle actuator driven by a stepper motor, the motion of which is regulated by the manual control. With the engine under automatic control an actuator is always required. The link between the actuator and the engine throttle lever must be free from backlash if good control is to be achieved. Some 'push–pull' cables have considerable internal backlash.

An important safety feature of a throttle actuator is that it should automatically move the engine control lever to the STOP position in the event of a power failure.

The role of the computer[2,5]

While there are still some circumstances in which it is appropriate to follow traditional procedures, with the operator controlling the test sequence manually and writing down the observed readings on a log sheet (the *training* of an engine test technician should certainly include practice in these procedures), more and more of this work is now handled by computers.

Many of the individual instruments or measuring systems forming part of a modern test cell have embedded computing capacity. These systems can communicate with other computers for the purposes of either data acquisition or control; care must be taken in the system design to ensure that the two-way communication between these devices is robust. It is common for a modern test cell to be controlled by, and reliant upon, a hierarchy of computers.

The tasks given to computers in engine testing may be considered under the following headings. These divisions are generally accepted in the industry, although some of the distinctions are becoming rather artificial as the technology develops.

Smart instrumentation and devices

Here the embedded intelligence is stored in some type of Read Only Memory (ROM) device. The programming of such devices is specific to some defined task and there is little or no reprogramming accessible to the user, apart from the use of defined routines such as calibration.

The interface between device and test cell computer may be by way of conventional analogue or digital I/O, by serial interface such as RS232 or IEEE or, increasingly, through a Local Area Network (LAN).

Typical devices in this category are gravimetric fuel consumption gauges and 'blow-by' meters. The top end of the category is represented by complex computer controlled systems such as those used to measure exhaust emissions. While such systems have considerable computing power it is 'inward looking' and largely inaccessible to (and unalterable by) the ordinary test bed user.

Engine management systems

These systems, a special type of device having substantial embedded intelligence, play a very important role in engine testing, indeed they are often themselves the subject of the test. Engine management systems are invariably designed to suit proprietary communication protocols.

The Engine Control Unit (ECU) can present problems in the test cell, particularly when, as is often the case, it includes means of immobilizing vehicles for security purposes; when this is so the system must be 'disarmed' or the matching security 'sender' device must be made available. Engine transducers must be integrated with the ECU, using wiring looms that may be special to the engine. There are at present attempts being made both in Europe and the USA to standardize communications with ECU's so that basic diagnostic systems may be standardized also.

Programmable Logic Controllers (PLC's)

These are used in the engine test industry for sequential control of engine handling operations such as automatic docking and for monitoring of complex interlocked services. They are considered to be better suited to processes requiring large amounts of digital switching rather than to rapid data acquisition and control, a task for which a conventional computer may be more appropriate.

Test bed computers

The test bed computer is considered by many engineers to be the most important piece of equipment in the control room: it is the means by which the engineer communicates with the whole test facility and it is therefore vital that it should be reliable and easy to use.

The TBC hardware often consists of a personal computer (PC) running one of the industrial standard operating systems: DOS, Windows NT, UNIX, etc. The authors' treatment, here and in Chapter 17, is based generally on the assumption that each cell will have its individual PC in a hierarchical position above any of the devices described above.

The test bed computer will contain much specialized application software which will vary in style between suppliers; however, all major suites contain software to interface with devices of the kinds listed above, together with engine throttle actuators and controls for whichever type of dynamometer is installed. In addition the suite will contain software for many different screen layouts, permitting the operator to interface with all aspects of the cell system. Such major software suites are the outcome of some man-years of work; it is not recommended that a user, however experienced, should undertake such a task from scratch.

Host computers

A host computer may range all the way from a PC joined to two or more cells up to a mainframe computer supervising the functioning of a complete factory or research facility. A host computer may play a number of roles: most frequently it is the main management tool of the test facility supervisor. It allows test sequences to be generated 'off line' and loaded down to individual cells and it may be used to create statistical summaries and to perform data reduction tasks. It is essentially a management tool since the computing power of the individual test bed computers is usually adequate.

Problems in achieving control

An engine test system is an assembly of an engine, a dynamometer and various actuators and peripherals. Each of these has its own control characteristics and, in many cases, its own controller. These controllers have not been designed as a group and it is not surprising that the combined control characteristics are often very far from ideal. The problem is compounded when the entire system is subjected to overall computerized control. At the simplest level, a dynamometer with a control loop intended to produce a particular torque–speed characteristic can generate instabilities quite absent in an old-fashioned manually controlled sluice-gate dynamometer. This gives a clue to dealing with control instabilities: eliminate as much of the control system as possible by switching to open-loop control.

It will also be found helpful, if two controllers are 'fighting' each other, to ensure that they have widely differing time-constants. In the common case of a speed-controlled engine coupled to a torque-controlled dynamometer it is preferable that the latter should have a shorter response time than the former.

Some engines are inherently difficult to control because of the shape of their torque–speed characteristics. Turbocharged diesel engines may have abrupt changes in the slope of the power–speed curve, plus sluggish response due to the time taken for speed changes in the turbocharger, which can make the optimization of the control system very difficult.

Inexperienced attempts to adjust a full three-term PID (proportional/ integral/differential) controller can lead to problems and it is well to record the settings before starting to make adjustments so that one can at least return to the starting point. Make sure that elementary sources of trouble, such as backlash and stiction in control linkages, are eliminated.

General characteristics of the electrical installation

Perhaps more than any other aspect of test cell design and construction the electrical installation is subject to regulations, most of which have statutory force. It is essential that any engineer responsible for the design or construction of a test cell should be thoroughly familiar, as a minimum, with the following:

- Electricity Supply Regulations of the Ministry of Fuel and Power
- BS 7671[11]: *Requirements for Electrical Installations: IEE Wiring Regulations*.

There are also many British Standards specifying individual features of an electrical system[1,4,7,8].

While these regulations cover most aspects of the electrical installation, and will not be further referred to here, there are several features that are a consequence of the special conditions associated with the test cell environment.

Fire precautions and the cable layout

It is wise to remember that fuel vapour may collect at floor level and seep into floor troughs. It is good practice to avoid installing any spark-inducing device less than 300 mm from floor level. Only in the most exceptional circumstances will it be necessary to specify 'explosion proof' electrical fittings: these can add considerably to the cost of the cell.

The floor channel ventilation system should be interlocked with the cell start-up procedure.

Where cables or cable trunking break through the test cell walls they must pass through a physical 'fire block'. There are some designs of cable wall-box that allow the fire block to be disassembled and extra cables added; none of these are particularly easy to use after a year or more in service and it is advisable to build in a number of spare cables.

Plastic 'soil pipes' cast into the floor are a convenient way of carrying signal cables between test cell and control room. These ducts should be laid to a fall in the direction of the cell and should have a raised lip to prevent drainage of liquid into them. Spare cables should be laid during installation. These pipes can be 'capped' by foam or filled with dried casting sand to create a noise, fire and vapour barrier.

Security of supply

It is important that the state of power supply systems should be indicated at the control desk. This is often overlooked and hours can be wasted in trying to reset computer systems when the cause of the trouble is that some service or peripheral device has been switched off. If, as is usually the case, the control and data acquisition system is computer based, it is sensible to include an uninterrupted power supply (UPS) in the specification. This will at least allow an orderly shut-down of the computer without loss of data.

Regenerative loading of electrical dynamometers

Under certain circumstances such as for example endurance testing, some of the long-running lubricant test procedures, Appendix II, and tests of large engines it may be worthwhile both economically and on ecological grounds to consider feedback of dynamometer power into the mains. There are invariably local regulations to be considered and the supply company must be advised. It may be necessary to install special protection devices to guard against possible damage or accidents should the mains trip out during regenerative operation.

Electrical interference

Interference[1] between power and signal circuits can be a major problem calling for expert consideration. Engine test cells are regions of high electromagnetic noise. High-voltage engine ignition systems are a major radiating source, while thyristor drives associated with electrical dynamometers give rise to harmonic distortion of the electrical supply and high-voltage 'spikes' that corrupt transducer and data processing signals.

Figure 3.4[9], is an example of a transducer signal corrupted by radiated noise. This is an example of burn-rate analysis, p.230, which is particularly sensitive to interference since it makes use of small differences between large pressure signals, thus increasing the noise to signal ratio.

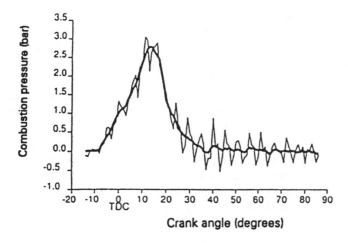

Figure 3.4. *Transducer signal corrupted by interference (the curve shows pressure rise due to combustion in a gasolene engine running at light load at 1500 rev/min, see p.233. It represents the difference between two pressure signals, one with and one without combustion, and thus accentuates interference effects)*

Transducer signals are usually 'conditioned' as near to the transducer as possible; nevertheless the resultant conditioned signals are commonly in the range 0 to 10 V d.c. or 0 to 5 mA, very small when compared with the voltage differences and current flows that may be present in power cables in the immediate vicinity of the signal lines.

The disturbance arising from a thyristor drive is a function of the relative rating of the machine and the supply transformer; clearly if the drive demands the full rated power of the transformer its capacity for causing distortion of the wave-form will be a maximum.

In general the susceptibility of the system to interference depends on the cable layout, and there are several steps that should be taken to minimize the problem.

- the mains supply to instrumentation should not be subject to distortion caused by other devices using the same sub-station distribution tapping. Equipment such as welding plant, variable speed drives and large motors should have their own separate supply.
- All instrument leads should be run in screened cable, preferably in trunking or flexible conduit. Parallel runs of power and signal lines should be avoided where possible and the distance between them kept to 0.3 m minimum.

- The design should minimize the length of sensitive circuits such as those of thermocouples, resistive temperature sensors and strain gauges prior to signal conditioning.
- The electrical specification must include details of earthing arrangements. Power units must have well defined, low-impedance earth circuit paths, quite separate from the instrumentation earths, otherwise 'earth loops' may arise, with stray currents finding their way into signal circuits. Instrument cable screening should also be earthed.
- Electromagnetic radiation in the radio frequency range, say from 100 kHz to 5 MHz, can be propagated over quite long distances and give rise to unpredictable disturbances in sensitive equipment. Switching of the drive current takes place typically at 300 Hz in d.c. thyristor drives and at up to 30 kHz in a.c. and servo drives; this can give rise to harmonics in the RF range.

European Safety Standards and CE marking

The European Community has, since 1985, been developing regulations to achieve technical harmonization and standards to permit free movement of goods within the Community. There are currently (early 1998) three directives of particular interest to the builders and operators of engine test facilities, (Table 3.1).

The use of the 'CE Mark' (abbreviation for 'Conformité Européen) implies that the manufacturer has complied with all directives issued by the EEC that are applicable to the product to which the mark is attached.

An engine test cell must be considered as the sum of many parts. Some of these parts will be items under test that may not meet the requirements of the relevant directives. Some parts will be standard electrical products that are able to carry their individual CE marks, while other equipment may range from

Table 3.1. *European Safety Standards and the CE Mark*

Reference	Directive	UK implementation
73/23/EEC with amendments	Low voltage	Electrical equipment (Safety) 1994
89/336/EEC with amendments	Electromagnetic compatibility	Electromagnetic compatibility 1992
89/392/EEC with amendments	Machinery	Supply of machinery (Safety) 1992

unique electronic modules to assemblies of products from various manufacturers. The situation is further complicated by the way in which electronic devices may be interconnected. If standard and tested looms join units belonging to a 'family' of products then the sum of the parts may comply with the relevant directive. If the interconnecting loom is unique to the particular plant the sum of the CE marked parts may not meet the strict requirements of the directive.

It is therefore not sensible for a specification for an engine test facility to include an unqualified global requirement that the facility 'be CE marked'. Some products are specifically excluded from the regulations while others are covered by their own rules; for example the directive 72/245/EEC covers radio interference from spark-ignition vehicle engines. Experimental and prototype engines may well fail to comply with this directive: an example of the impossibility of making any unqualified commitment to comply in all respects at all times with these bureaucratic regulations.

The reader is advised to consult specific 'Health and Safety' literature, or that produced by Trade Associations, if in doubt regarding the way in which these directives should be treated.

Heat: a further source of instrument error

Instrument errors caused by heat are particularly difficult to trace, as the instrument will probably be calibrated when cold, so they should be eliminated at source. Many instrumentation packages produce quite appreciable quantities of heat and if mounted low down in the confined space of a standard 19-in-rack cabinet they may raise the temperature of apparatus mounted above them over the generally specified maximum of 40°C. Control and instrument cabinets should be well ventilated, and it may be necessary to supplement individual ventilation fans by extraction fans high in the cabinet. If dust is present the cooling air should be filtered at inlet.

Special attention should be given to heat insulation and ventilation when instruments are carried on an overhead boom.

It is good practice to fit all instrument housings with simple (magnetically attached) temperature indicators.

Summary: the test cell specification

The first three chapters of this book have touched upon a large number of complex matters, all of which need to be considered by any engineer who is given the task of drawing up the specification for a new test facility.

As in all complex tasks, the first step should be to draw up a concise statement of the *purposes* the installation is intended to serve. This is not necessarily an easy undertaking, since within the organization there will be a number of internal 'customers' with conflicting interests that need to be reconciled.

If the cell is intended for research and development the engineering department will undoubtedly press for the most elaborate installation that has a chance of securing financial approval. There is a particular temptation to build in facilities that may (or may not) be needed in the future. If the required maximum speed of a 100 kW electrical dynamometer is limited to 8500 rev/min there are several sources of supply at reasonable prices. If the speed is raised to 10 000 rev/min, in anticipation of possible future increases in engine speed, there are fewer machines to choose from, at much higher prices.

It is always sound policy to find out what is available on the market at an early stage, and to reconsider carefully any part of the specification that makes demands that exceed what is commonly offered.

If the installation is intended for routine testing – endurance, quality control or production – it is important to draw the right balance between the manpower costs involved in running the installation and the investment on automation, engine handling and rigging equipment. This depends very much on the number of engine changes per day or shift and the load factor envisaged for the cell.

The major components of a test cell specification are:

- statement of objectives of the installation
- overall description of cell space and control room area (this often means adaptation of an existing facility)
- choice of dynamometer and control system
- engine handling and rigging equipment
- range of instrumentation to be provided
- data acquisition, data processing and control equipment
- test cell services: fuel, electrical power, ventilation, water, exhaust, fire control and safety
- specification of acceptance tests, handover procedure and training.

A note on documentation

Test cells and control rooms are in the nature of things subject to modification. The initial documentation, representing the 'as commissioned' state of the facility, must be of a high standard and easily accessible to maintenance staff and contractors. Responsibility for keeping records and schematics up to date must be clearly defined, see Chapter 18. Originals should be in a reproducible form.

Any contract for a new or modified facility should specify the supply of 'as commissioned' drawings.

References

1. BS 6667 Part **3** *Method of Evaluating Susceptibility to Radiated Electromagnetic Energy.*
2. Grantham, W.J. and Vincent, T.L. (1993) *Modern Control Systems Analysis and Design,* Wiley, Chichester.
3. BS 6206 *Specification for impact Performance Requirements for Flat Safety Glass and Safety Plastics for use in Buildings.*
4. BS 6739 *Code of Practice for Instrumentation in Process Control Systems: Installation Design and Practice.*
5. Mahoud, M.S. (1991) *Computer-operated Systems Control,* Marcel Dekker, New York.
6. Parr, E.A. (1998) *Industrial Control Handbook*, Butterworth-Heinemann, Oxford.
7. BS 4293 *Specification for Residual Current-operated Circuit-breakers.*
8. BS 4678 Parts 1, 2, 4, *Cable Trunking.*
9. Shayler, P.J. *et al.* (1990) *Improving the Determination of Mass Burn Fraction,* S.A.E. Paper No. 900351
10. Leigh, J.R. (1985) *Applied Digital Control,* Prentice-Hall, London.
11. BS 7671 *Requirements for Electrical Installations: IEE Wiring Regulations* (16th Edition).

4 Ventilation and air conditioning

In this chapter the concept of the test cell as an open system is applied to the analysis of thermal loadings and ventilation requirements[1,2]. The design process for ventilation systems is described with a worked example. It must be remembered that an i.c. engine is essentially an air engine, see Chapter 10. The air used by the engine may come from the ventilation air or from outside the cell. Either way the performance and power output of the engine are affected by the condition, temperature, pressure and humidity of the air.

The purpose of air conditioning and ventilation is the maintenance of an acceptable environment in an enclosed space. This is a comparatively simple matter where only human activity is taking place, but becomes progressively more difficult as the energy flows into and out of the space increase. An engine test cell represents perhaps the most demanding environment encountered in industry. Large amounts of power will be generated in a comparatively small space, surfaces at high temperature are unavoidable and large flows of cooling water, air and electrical power have to be accommodated, together with rapid variations in load.

The heat capacity of cooling air

By definition, the test cell environment is mainly controlled by regulating the quantity, temperature and in some cases the humidity of the air passing through it. Air is not the ideal heat transfer medium: it has low density and specific heat and is transparent to radiant heat, while its ability to cool hot surfaces is much inferior to that of liquids.

The main properties of air of significance in air conditioning may be summarized as follows:

the gas equation:

$$p_a \times 10^5 = \rho R(t_a + 273) \tag{1}$$

where:

p_a = atmospheric pressure, bar
ρ = air density, kg/m^3
R = gas constant for air = 287 J/kg K
t_a = air temperature, °C

Under conditions typical of test cell operation, with $t_a = 25$ (77°F), and standard atmospheric pressure (see Units and Conversion Factors), the density of air, from eq. (1), is

$$\rho = \frac{1.01325 \times 10^5}{287 \times 298} = 1.185 \, \text{kg/m}^3$$

or about 1/850th that of water.

The specific heat at constant pressure of air at normal atmospheric conditions is approximately:

$$C_p = 1.01 \, \text{kJ/kg K}$$

or less than one quarter that of water.

The air flow necessary to carry away 1 kW of power with a temperature rise of 10°C is:

$$m = \frac{1}{1.01 \times 10} = 0.099 \, \text{kg/s} = 0.084 \, \text{m}^3/\text{s} \, (2.9 \, \text{ft}^3/\text{s})$$

This is a better basis for design than any rule of thumb regarding number of cell air changes per hour.

Heat transfer from the engine

It is useful to gain a feel for the relative significance of the elements that make up the total of heat transferred from a running engine to its surroundings by considering rates of heat transfer from bodies of simplified form under test cell conditions[3].

Consider a body of the shape sketched in Fig. 4.1. This might be regarded as roughly equivalent, in terms of projected surface areas in horizontal and vertical directions, to a gasolene engine of perhaps 100 kW maximum power output although the total surface area of the engine would be much greater. Let

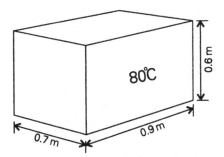

Figure 4.1. *Simplified model of 100 kW engine, for analysis of heat transfer to surroundings*

us assume the surface temperature of the body to be 80°C and the temperature of the cell air and cell walls 30°C.

Heat loss occurs as a result of two mechanisms: natural convection and radiation. The rate of heat loss by natural convection from a vertical surface in *still air* is given approximately[3] by:

$$Q_v = 1.9(t_s - t_a)^{1.25} \text{ W/m}^2 \tag{2}$$

The total area of the vertical surfaces in Fig. 4.1 = 1.9 m². The corresponding convective loss is therefore:

$$1.9 \times 1.9 \times (80 - 30)^{1.25} = 480 \text{ W}$$

The rate of heat loss from an upward facing horizontal surface is approximately:

$$Q_h = 2.5(t_s - t_a)^{1.25} \tag{3}$$

giving in the present case a convective loss of:

$$0.63 \times 2.5(80 - 30)^{1.25} = 210 \text{ W}$$

The heat loss from a downward facing surface is about half that for the upward facing case, giving a loss = 110 W.

We thus arrive at a rough estimate of convective loss of 800 W. For a surface temperature of 100°C this would increase to about 1200W.

However, this is the heat loss in still air: the air in an engine test cell is anything but still, and very much greater rates of heat loss can and probably will occur. *This effect must never be forgotten in considering cooling problems in a test cell: an increase in air velocity greatly increases the rate of heat transfer to the air and may thus aggravate the problem.*

As a rough guide, doubling the velocity of air flow past a hot surface increases the heat loss by about 50%. The air velocity due to natural convection in our example is about 0.3 m/s.

An air velocity of 3 m/s would be moderate for a test cell with ventilating fans producing a vigorous circulation, and such a velocity past the body of Fig. 4.1 would increase convective heat loss fourfold, to about 3.2 kW at 80°C, 4.8 kW at 100°C.

The rate of heat loss by radiation from a surface depends on the *emissivity* of the surface (the ratio of the energy emitted to that emitted by a so-called *black body* of the same dimensions and temperature) and on the temperature difference between the body and its surroundings[3]. Air is essentially transparent to radiation, which thus serves mainly to heat up the surfaces of the surrounding cell; this heat must subsequently be transferred to the cooling air by convection, or conducted to the surroundings of the cell.

Heat transfer by radiation is described by the Stefan–Boltzmann equation, a convenient form of which is:

$$Q_r = 5.77\varepsilon\left[\left(\frac{t_s + 273}{100}\right)^4 - \left(\frac{t_w + 273}{100}\right)^4\right] \qquad (4)$$

A typical value of emissivity (ε) for machinery surfaces would be $= 0.9$. $t_s =$ temperature of hot body (°C) and $t_w =$ temperature of enclosing surface (°C). In the present case:

$$Q_r = 5.77 \times 0.9\left[\left(\frac{353}{100}\right)^4 - \left(\frac{303}{100}\right)^4\right] = 370 \text{ W/m}^2$$

In our present example total surface area $= 3.16 \text{ m}^2$, giving a radiation heat loss of 1170 W.

For a surface temperature of 100°C this would increase to 1800 W.

Heat transfer from the exhaust system

The other main source of heat loss associated with the engine is the exhaust system. In the case of turbocharged engines this can be particularly significant. Assume in the present example that exhaust manifold and exposed exhaust pipe are equivalent to a cylinder 80 mm dia × 1.2 m long at a temperature of 600°C, surface area 0.3 m², Fig. 4.2. Heat loss at this high temperature will be predominantly by radiation and equal to:

$$0.3 \times 5.77 \times 0.9\left[\left(\frac{873}{100}\right)^4 - \left(\frac{303}{100}\right)^4\right] = 8900 \text{ W}$$

from eq. (4)
Convective loss is:

$$0.3 \times 1.9 \times (600 - 30)^{1.25} = 1600 \text{ W}$$

from eq. (2).

It is clear that this can heavily outweigh the losses from the engine, and points to the importance of reducing the run of unlagged exhaust pipe as much as possible.

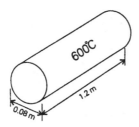

Figure 4.2. *Simplified model of 100 kW engine exhaust manifold*

Heat losses from the bodies sketched in Figs. 4.1 and 4.2, representing an engine and exhaust system, surroundings at 30°C, may be summarized as follows:

	Convection	*Radiation*
engine, jacket and crankcase at 80°C	3.2 kW	1.2 kW
engine, jacket and crankcase at 100°C	4.8 kW	1.8 kW
exhaust manifold and tailpipe at 600°C	1.6 kW	8.9 kW

Heat transfer from walls

Most of the heat radiated from engine and exhaust system will be absorbed by the cell walls and ceiling, also by instrument cabinets and control boxes, and subsequently transferred to the ventilation air by convection.

Imagine the 'engine' and 'manifold' considered above to be installed in a test cell of the dimensions shown in Fig. 4.3. The total wall area is 88 m². Assuming a wall temperature 10°C higher than the mean air temperature in the cell and an air velocity of 3 m/s the rate of heat transfer from wall to air is in the region of 100W/m², or 8.8 kW for the whole wall surface, roughly 90% of the heat radiated from engine and exhaust system; the equilibrium wall temperature is perhaps 15°C higher than that of the air.

While an attempt to make a detailed analysis on these lines, using exact values of surface areas and temperatures, would not be worth while, this simplified treatment may clarify the principles involved.

Sources of heat in the test cell

The engine

Various estimates of the total heat release to the surroundings from a water

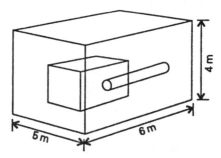

Figure 4.3. *Simplified test cells for heat transfer calculations*

cooled engine and its exhaust system have been published. One authority[4] quotes a maximum of 15% of the heat energy in the fuel, divided equally between convection and radiation. This would correspond to about 30% of the power output of a diesel engine and 40% in the case of a gasolene engine.

In the experience of the authors a figure of 40% (0.4 kW/kW engine output) represents a safe upper limit to be used as a basis for design for water-cooled engines. This is divided roughly in the proportion 0.1 kW/kW engine to 0.3 kW/ kW exhaust system. It is thus quite sensitive to exhaust layout and insulation.

In the case of an air-cooled engine the heat release from the engine will increase to about 0.7 kW/kW output in the case of a diesel engine and to about 0.9 kW/kW output for a gasolene engine. The proportion of the heat of combustion that passes to the cooling water in a water-cooled engine, in an air-cooled unit must of course pass directly to the surroundings.

The dynamometer

A water-cooled dynamometer, whether hydraulic or eddy current, runs at a moderate temperature and heat losses to the cell are unlikely to exceed 5% of power input to the brake. Usually a.c. or d.c. machines are air-cooled, and heat loss to surroundings is in the region of 15% of power input.

Other sources of heat

Effectively all the electrical power to lights, fans and instrumentation in the test cell will eventually appear as heat transmitted to the ventilation air. The same applies to the power taken to drive the forced-draught ventilating fans: this is dissipated as heat in the air handled by the fans.

Heat losses from the cell

The temperature in an engine test cell is generally higher than usual for an industrial environment. There is thus in some cases appreciable transfer of heat through cell walls and ceiling, depending on the configuration of the site but, except in the case of a test cell forming an isolated unit, these losses may probably be neglected.

Recommended values as a basis for the design of the ventilation system are given in Table 4.1. In all cases they refer to the maximum rated power output of the engines to be installed.

Calculation of ventilation load

The first step is to estimate the various contributions to the heat load from engine, exhaust system, dynamometer, lights and services. This information

Table 4.1.   *Heat transfer to ventilation air*

	kW/kW power output
Engine, water cooled	0.1
Engine, air cooled	0.7–0.9
Exhaust system (manifold and silencer)	0.3
Hydraulic dynamometer	0.05
Eddy current dynamometer	0.05
a.c./d.c. dynamometer	0.15

should be summarized in a single flow diagram, see Chapter 6. Table 4.1 shows typical values.

Heat transfer to the ventilation air is to a degree self-regulating: the cell temperature will rise to a level at which there is an equilibrium between heat released and heat carried away. The amount of heat carried away by a given air flow is clearly a function of the temperature rise ΔT from inlet to outlet.

If the total heat load is H_L kW then the required air flow rate is:*

$$Q_A = \frac{H_L}{1.01 \times 1.185 \, \Delta T} = 0.84 \frac{H_L}{\Delta T} \; \mathrm{m^3/s} \tag{5}$$

A temperature rise $\Delta T = 10°C$ is a reasonable basis for design. Clearly the higher the value of ΔT the smaller the corresponding air flow. However, a reduction in air flow has two influences on general cell temperature: the higher the outlet temperature the higher the mean level in the cell, while a smaller air flow implies lower air velocities in the cell, calling for a greater temperature difference between cell surfaces and air for a given rate of heat transfer.

Design of ventilation ducts and distribution systems

Pressure losses: fundamentals

The velocity head or pressure associated with air flowing at velocity V is given by:

$$p_v = \frac{\rho V^2}{2} \, \mathrm{Pa}$$

This represents the pressure necessary to generate the velocity. A typical value for ρ, density of air, is $1.2 \, \mathrm{kg/m^3}$, giving:

---

*For air at standard atmospheric conditions, $\rho = 1.185$ kg/m^3. A correction may be applied when density departs from this value but is probably hardly worth while.

$$p_v = 0.6V^2 \text{ Pa}$$

The pressure loss per metre length of a straight duct is a fairly complex function of air velocity, duct cross-section and surface roughness. Methods of derivation with charts are given in ref. 5. In general the loss lies within the range 1–10 Pa/m, the larger values corresponding to smaller duct sizes. For test cells with individual ventilating systems duct lengths are usually short and these losses are small compared with those due to bends and fittings.

The choice of duct velocity is a compromise depending on considerations of size of ducting, power loss and noise. If design air velocity is doubled the size of the ducting is clearly reduced but tbe pressure losses are increased roughly fourfold, while the noise level is greatly increased (by about 18 dB for a doubling of velocity).

Maximum duct velocities recommended in ref. 5 are given in Table 4.2. The total pressure of an air flow P_t is the sum ofthe velocity pressure and the static pressure p_s (relative to atmosphere):

$$p_t = p_s + p_v = p_s + \frac{\rho V^2}{2} \tag{6}$$

The design process for a ventilating system includes the summation of the various pressure losses associated with the different components and the choice of a suitable fan to develop the total pressure required to drive the air through the system.

Ducting and fittings

Various Codes of Practice have been produced covering the design of ventilation systems[6,7]. The following brief notes are based on the *Design Notes* published by the Chartered Institution of Building Services[5].

Galvanized sheet steel is the most commonly used material, and ducting is readily available in a range of standard sizes in rectangular or circular sections. Rectangular section ducting has certain advantages: it can be fitted against flat

Table 4.2. *Maximum recommended duct velocities*

Volume rate of flow (m³/s)	Maximum velocity (m/s)	Velocity pressure (Pa)
<0.1	8–9	38–55
0.1–0.5	9–11	55–73
0.5–1.5	11–15	73–135
> 1.5	15–20	135–240

Pressures in ventilating systems are often measured in mm H_2O:
1 mm $H_2O = 9.81$ Pa.

surfaces and expensive round-to-square transition lengths for connection to components of rectangular section such as centrifugal fan discharge flanges, filters and coolers are avoided.

Once the required air flow rate has been settled and the general run of the ducting decided the next step is to calculate the pressure losses in the various elements in order to specify the pressure to be developed by the fan. In most cases a cell will require both a forced draught fan for air supply and an induced draught fan to extract the air. The two fans must be matched to maintain the cell pressure as near as possible to atmospheric.

Figure. 4.4 shows various components in diagrammatic form and indicates the loss in total pressure associated with each. This loss is given by:

$$\Delta p_t = K_e \frac{\rho V^2}{2} \tag{7}$$

Information on pressure losses in plant items such as filters, heaters and coolers is generally provided by the manufacturer.

The various losses are added together to give the cumulative loss in total

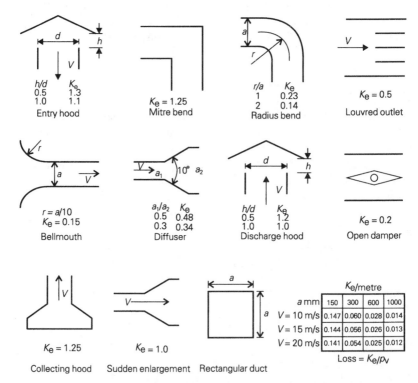

Figure 4.4.   *Pressure losses in components of ventilation system*

pressure (static pressure + velocity pressure) which determines the required fan performance.

Inlet and outlet ducting

The arrangement of air inlets and outlets calls for careful consideration if short-circuiting and local areas of stagnant air are to be avoided. Possibilities are:

- inlet louvres at low level with no inlet fan. High-level outlet with extraction fans. System prone to 'dead zones'.
- high-level inlet duct with inlet fans and directional air flow. Low-level extraction duct at opposite end of cell. Probably the best arrangement.
- the reverse of the previous arrangement with low-level inlet and high-level outlet.

The choice of system has a major influence on the layout of equipment in the cell. The authors do not greatly favour an extraction hood above the engine, which may make engine handling difficult and also affect engine inlet air conditions.

The use of 'spot fans' for supplementary cooling

The indiscriminate use of high-velocity fans to deal with local hotspots can greatly increase the total amount of heat transferred to the ventilation air, with a consequent increase in overall cell temperature. A jet of hot air can have undesirable effects: for example it can raise the temperature of the engine inlet air and upset the calibration of force transducers.

Fans

Methods of testing fans and definitions of fan performance are given in BS 848[8]. This is not an entirely straightforward matter; a brief summary follows. Once again the control volume technique will be found useful. Figure 4.5 shows a fan surrounded by a control surface. The various flows into and out of the control volume are as follows.

In air flow Q_F at velocity V_1 and pressure p_{s_1} power input P_A
Out air flow Q_F at velocity V_2 and pressure p_{s_2} where p_{s_1} and p_{s_2} are the static pressures at inlet and outlet respectively.

Total pressure at inlet

$$p_{t_1} = p_{s_1} + \frac{\rho V_1^2}{2} \tag{8}$$

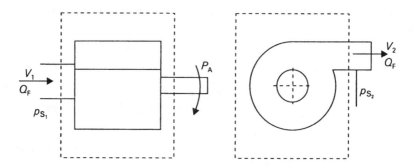

Figure 4.5.   *Centrifugal fan as an open system*

Total pressure at outlet

$$p_{t_2} = p_{s_2} + \frac{\rho V_2^2}{2} \tag{9}$$

Fan total pressure

$$p_{t_F} = p_{t_2} - p_{t_1} \tag{10}$$

Air power (total)

$$P_{t_F} = Q_F P_{t_F} \tag{11}$$

Most fan manufacturers quote *fan static pressure* which is defined as:

$$p_{s_F} = p_{t_F} - \frac{\rho V_2^2}{2} \tag{12}$$

this ignores the velocity pressure of the air leaving the fan and does *not* equal the pressure difference $p_{s_2} - p_{s_1}$ between the inlet and outlet static pressures.

(It is worth noting that, in the case of an axial flow fan with free inlet and outlet, fan static pressure as defined above will be zero.)

The total air power P_{t_F} is a measure of the power required to drive the fan in the absence of losses.

The air power (static) is given by:

$$P_{s_F} = Q_F p_{s_F} \tag{13}$$

Manufacturers quote either fan static efficiency η_{s_A} or fan total efficiency η_{t_A} The shaft power is given by:

$$P_A = \frac{P_{s_F}}{\eta_{s_A}} = \frac{P_{t_F}}{\eta_{t_A}} \tag{14}$$

Fan noise

In a ventilation system the fan is usually the main source of system noise. The noise generated varies as the square of the fan pressure head so that doubling the system resistance for a given flow rate will increase the fan sound power fourfold, or by about 6 dB.

As a general rule, to minimize noise, ventilation fans should operate as close to the design point as possible.

Ref. 6 gives examples of good and bad designs of fan inlet and discharge; a poor design gives rise to increased noise generation. Ref. 8, Part 2, gives guidance on methods of noise testing fans.

Classification of fans[8]

1. *Axial flow fans.* For a given flow rate an axial flow fan is considerably more compact than the corresponding centrifugal fan, and fits very conveniently into a duct of circular cross-section. The fan static pressure per stage is limited, typically to a maximum of about 600 Pa at the design point, while the fan dynamic pressure is about 70% of the total pressure. Fan total efficiencies are in the range 65–75%. An axial flow fan is a good choice as a spot fan, or for mounting in a cell wall without ducting. Multi-stage units are available, but tend to be fairly expensive.

2. *Centrifugal fans, flat blades, backward inclined.* This is probably the first choice in most cases where a reasonably high pressure is required, as the construction is cheap and efficiencies of up to 80% (static) and 83% (total) are attainable. A particular advantage is the immunity of the flat blade to dust collection. Maximum pressures are in the range 1–2 kPa.

3. *Centrifugal fans, backward curved.* These fans are more expensive to build than the flat bladed type. Maximum attainable efficiencies are 2–3% higher, but the fan must run faster for a given pressure and dust tends to accumulate on the concave faces of the blades.

4. *Centrifugal fans, aerofoil blades.* These fans are expensive and sensitive to dust, but are capable of total efficiencies exceeding 90%. There is a possibility of discontinuities in the pressure curve due to stall at reduced flow. They should be considered in the larger sizes where the savings in power cost are significant.

5. *Centrifugal fans, forward curved blades.* These fans are capable of a delivery rate up to $2\frac{1}{2}$ times as great as that from a backward inclined fan of the same size, but at the cost of lower efficiency, unlikely to exceed 70% total. The power curve rises steeply if flow exceeds the design value.

The various advantages and disadvantages are summarized in Table 4.3.

Table 4.3. *Fans: advantages and disadvantages*

Fan type	Advantages	Disadvantages
Axial flow	Compact Convenient installation Useful as free-standing units	Moderate efficiency Limited pressure Fairly expensive
Centrifugal, flat blades, backward inclined	Cheap Capable of high pressure Good efficiency Insensitive to dust	
Centrifugal, backward curved	Higher efficiency	More expensive Higher speed for given pressure Sensitive to dust
Centrifugal aerofoil	Very high efficiency	Expensive Sensitive to dust May stall
Centrifugal forward curved	Small size for given duty	Low efficiency Possibility of overload

Design of ventilation system: worked example

By way of illustration, consider the case of the 250 kW turbocharged diesel engine for which an energy balance is given in Chapter 11, the engine to be coupled to a hydraulic dynamometer.

Ventilation airflow

Assume convection and radiation losses as follows:

engine, 10% of power output	25 kW
exhaust manifold	15 kW
exhaust tailpipe and silencer	15 kW
dynamometer, 5% of power input	12 kW
lights and services	20 kW
forced draught fan	5 kW
	92 kW
less losses from cell walls by conduction	5 kW
Total	87 kW

Assume air inlet temperature 20°C, outlet temperature 31°C, $\Delta T = 11°C$. Then air flow rate, from eq. (5) is:

$$Q_A = \frac{0.84 \times 87}{11} = 6.6 \, \text{m}^3/\text{s}$$

+ induction air, 0.3 m³/s, say 7 m³/s (= 101 m³/h per kW engine power output). Assuming cell dimensions 8 m × 6 m × 4.5 m high, cell volume = 216 m³, this gives 117 air changes per hour.

Table 4.2 suggests a mean duct velocity in the range 15–20 m/s as appropriate, giving a cross-sectional area of 0.37–0.49 m². Heinsohn[2] gives a range of recommended standard duct dimensions of which the most suitable in the present circumstances is 600 mm × 600 mm, giving a duct velocity of 19.5 m/s and a velocity pressure of 228 Pa.

Figure 4.6 shows a possible layout for the ventilation system. The inlet or forced draught system uses a centrifugal fan and the duct velocity assumed above. For the extraction system, with its simpler layout and smaller pressure losses, an axial flow fan has been chosen. The pressure losses are calculated in Table 4.4 (this process lends itself readily to computer programming).

Table 4.4 indicates fan duties as follows:
forced draught, flow rate 7 m³/s, static pressure 454 Pa; extraction, flow rate 7 m³/s, static pressure 112 Pa.
A manufacturer's catalogue offers a fan for the forced draught situation to the following specification:

Centrifugal, backward inclined	
Impeller diameter	900 mm
Speed	850 rev/min
Fan static efficiency	65%

giving, from eqns (13) and (14)

$$\text{shaft power} = \frac{7.0 \times 454}{0.65} = 4900 \, \text{W}$$

$$\text{motor power} = 5 \, \text{kW}$$

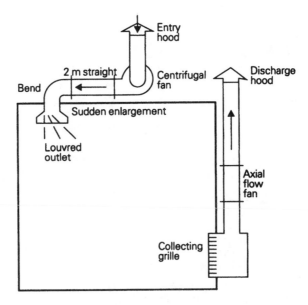

Figure 4.6.   *Layout of test cell ventilation system*

For the extraction fan, an axial flow unit has the following specification:

Diameter   1 m
Speed   960 rev/min
Velocity pressure 51 Pa
Fan total efficiency 80%

$$\text{shaft power} = \frac{7.0 \times (112 + 51)}{0.80} = 1400 \text{ W}$$

In this case the manufacturer recommends a motor rated at 2.2 kW.

Ventilation of the control room

This is in general a much less demanding exercise for the test installation designer. Heating loads are moderate, primarily associated with lights and heat generated by electronic apparatus located in the room. Regulations regarding air flow rate per occupant and general conditions of temperature and humidity are laid down in various codes of practice[6,7], and should be equivalent to those considered appropriate for offices.

Air conditioning

Most people associate this topic with comfort levels under various conditions: sitting, office work, manual work, etc. Levels of air temperature and humidity,

Table 4.4. *Calculation of system pressure loss*

Item	Size (mm)	Area (m²)	Volume Flow rate (m³/s)	Velocity (m/s)	Velocity pressure (Pa)	Fitting loss factor K_e	Pressure drop K_e/m	Length (m)	Pressure loss (Pa)	Cumulative loss (Pa)
Forced draught										
Entry hood	800 dia	0.5	7.0	14	118	1.1			129	129
Straight	800 dia	0.5	7.0	14	118		0.02	2	5	134
Bend	800 dia	0.5	7.0	14	118	0.23			27	161
Fan										
Straight	600 × 600	0.36	7.0	19	217		0.025	2	11	172
Bend	600 × 600	0.36	7.0	19	217	0.23			50	222
Sudden enlargement	600 × 600	0.36	7.0	19	217	1.0			217	439
Louvred outlet	1000 × 1000	1.00	7.0	7	29	0.5			15	454
								Total	454	
Extraction										
Collecting hood	1000 dia*	0.79	7.0	9	49	1.25			61	61
Straights	1000 dia	0.79	7.0	9	49		0.02	2	2	63
Fan										
Discharge hood	1000 dia	0.79	7.0	9	49	1.0			49	112
								Total	112	

*Diameter of branch connection.

also air change rates are laid down by statute. See BS 5720[7] for particulars. Such regulations must be observed with regard to the control room. However, the conditions in an engine test cell are far removed from the normal, and justify special treatment, which follows.

Two properties of the ventilating air entering the cell (and, more particularly, of the induction air entering the engine) are of importance: the temperature and the moisture content. Air conditioning involves four main processes:

- heating the air
- cooling the air
- reducing the moisture content (dehumidifying)
- increasing the moisture content (humidifying).

Fundamentals of psychrometry

The study of the properties of moist air is known as psychrometry. It is treated in many standard texts[1,2] and only a very brief summary will be given here.

Air conditioning processes are represented on the psychrometric chart, Fig. 4.7[9] This relates the following properties of moist air:

- the *moisture content or specific humidity,* ω kg moisture/kg dry air. Note that even under fairly extreme conditions (saturated air at 30°C) the moisture content does not exceed 3% by weight
- the *percentage saturation* or *relative humidity,* ϕ. This is the ratio of the mass of water vapour present to the mass that would be present if the air were saturated at the same conditions of temperature and pressure. The mass of vapour under saturated conditions is very sensitive to temperature:

Temperature (°C)	10	15	20	25	30
Moisture content ω(kg/kg)	0.0076	0.0106	0.0147	0.0201	0.0273

A consequence of this relationship is the possibility of drying air by cooling. As the temperature is lowered the percentage saturation increases until at the *dew point temperature* the air is fully saturated and any further cooling results in the deposition of moisture.

- The *wet- and dry-bulb temperatures.* The simplest method of measuring relative humidity is by means of a wet- and dry-bulb thermometer. If unsaturated air flows past a thermometer having a wetted sleeve of cotton around the bulb the temperature registered will be less than the actual temperature of the air, as registered by the dry-bulb thermometer, owing to evaporation from the wetted sleeve. The difference between the wet-bulb and dry-bulb temperatures is a measure of the relative humidity. Under

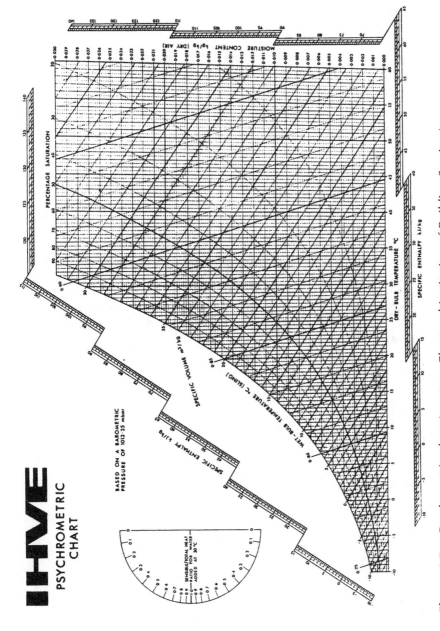

Figure 4.7. *Psychrometric chart (courtesy Chartered Institution of Building Services).*

saturated conditions the temperatures are identical, and the depression of the wet-bulb reading increases with increasing dryness. Wet- and dry-bulb temperatures are shown in Fig. 4.7.

● The *specific enthalpy* of the air, relative to an arbitrary zero corresponding to dry air at 0°C. We have seen earlier (p.47) that on this basis the specific enthalpy of dry air

$$h = C_p t_a = 1.01 t_a \text{ kJ/kg} \tag{15a}$$

The specific enthalpy of *moist* air must include both the sensible heat and the *latent heat of evaporation* of the moisture content.

The specific enthalpy of moist air:

$$h = 1.01 t_a + \omega(1.86 t_a + 2500) \tag{15b}$$

the last two terms representing the sum of the sensible and latent heats of the moisture.

Taking the example of saturated air at 30°C:

$$h = 1.01 \times 30 + 0.0273(1.86 \times 30 + 2500)$$

$$= 30.3 + 1.5 + 68.3 \text{ kJ/kg}$$

The first two terms represent the sensible heat of air plus moisture, and it is apparent that ignoring the sensible heat of the latter, as is usual in air cooling calculations, introduces no serious error. The third term, however, representing the latent heat of the moisture content, is much larger than the sensible heat terms. This accounts for the heavy cooling load associated with the process of drying air by cooling: condensation of the moisture in the air is accompanied by a massive release of latent heat.

Air conditioning processes

Heating and cooling without deposition of moisture

The simple expression for dry air:

$$H = \rho Q_A C_p \Delta T \text{ kW} \tag{16}$$

is adequate.

Cooling to reduce moisture content

This is a very energy-intensive process, and is best illustrated by a worked example.

Increasing moisture content This is achieved either by spraying water into the air stream (with a corresponding cooling effect) or by steam injection. *With the advent of Legionnaires' disease the latter method, involving steam that is essentially sterile, is favoured.*

Calculation of cooling load: worked example Consider the ventilation system described above. We have assumed an air flow rate $Q_A = 7\,\text{m}^3/\text{s}$, air entering at 20°C. If ambient temperature is 25°C and we are required to reduce this to 20°C, then from eq. (16) assuming saturation is not reached,

$$\text{cooling load} = 1.2 \times 7.0 \times 1.01 \times 5 = 42.4\,\text{kW}$$

Now let us assume that the ambient air is 85% saturated, $\phi = 0.85$ and that we need to reduce the relative humidity to 50% at 20°C.

The psychrometric chart shows that the initial conditions correspond to a moisture content:

$$\omega_1 = 0.0171\,\text{kg/kg}$$

The final condition, after cooling and dehumidifying corresponds to:

$$\omega_2 = 0.0074\,\text{kg/kg}$$

The chart shows that the moisture content ω_2 corresponds to saturation at a temperature of 9.5°C.

From eq. (15b) the corresponding specific enthalpies are

$$h_1 = 1.01 \times 25 + 0.0171(1.86 \times 25 + 2500)$$

$$= 68.80\,\text{kJ/kg}$$

$$h_2 = 1.01 \times 9.5 + 0.0074(1.86 \times 9.5 + 2500)$$

$$= 28.22\,\text{kJ/kg}$$

The corresponding cooling load

$$L_C = 1.2 \times 7.0 \times (68.80 - 28.22)$$

$$= 341\,\text{kW}$$

If it is required to warm the air up to the desired inlet temperature of 20°C the specific enthalpy is increased to:

$$h_3 = 1.01 \times 20 + 0.0074(1.86 \times 20 + 2500)$$

$$= 39.00\,\text{kJ/kg}$$

The corresponding heating load:

$$L_H = 1.2 \times 7.0 \times (39.00 - 28.22)$$

$$= 90.5\,\text{kW}$$

To summarize:

Air flow of $7 \, \text{m}^3/\text{s}$ at 25°C, 85% saturated

Cooling load to reduce temperature to 20°C	42.4 kW
Cooling load to reduce temperature to 9.5°C with dehumidification.	341 kW
Heating load to restore temperature to 20°C, 50% saturated.	90.5 kW

Cooling of air is usually accomplished by heat exchangers fed with chilled water. See ref. (10) for a description of liquid chilling packages.

This illustrates the fact that any attempt to reduce humidity by cooling as opposed to merely reducing the air temperature without reaching saturation conditions imposes very heavy cooling loads.

Calculation of humidification load: worked example   Let us assume initial conditions, $Q_A = 7 \, \text{m}^3/\text{s}$, temperature 20°C, relative humidity $\phi = 0.3$, and that we are required to increase this to $\phi = 0.7$ for some experimental purpose. Relative moisture contents are:

$$\omega_1 = 0.0044 \, \text{kg/kg}$$

$$\omega_2 = 0.0104 \, \text{kg/kg}$$

Then rate of addition of moisture

$$= 7 \times 1.2 \times (0.0104 - 0.0044) = 0.050 \, \text{kg/s}$$

Taking the latent heat of steam as 2500 kJ/kg,

$$\text{heat input} = 2500 \times 0.050 = 125 \, \text{kW}$$

This is again a very large load, calling for a boiler of at least this capacity.

Conditioning of combustion air

It is in general a practical proposition only to manipulate the relative humidity of the air entering the engine, leaving the condition of the cell ventilation air unchanged. In the present example this would amount to an air flow of $0.24 \, \text{m}^3/\text{s}$, corresponding to a reduction in the cooling and heating loads calculated above in the ratio of 1 : 29, a much more reasonable proposition.

While the test engineer is unlikely to be directly involved in the design of complete cell air conditioning systems, usually the province of the specialist, he may be required to design the conditioning system for the combustion air supply to an engine. The condition, pressure, temperature, moisture content and purity of the combustion air has a number of significant effects on engine performance; these are described in Chapter 10, p.183 while the actual design of a conditioning system for combustion air is described in Chapter 6, p.93

Effects of humidity: a warning

Electronic equipment is extremely sensitive to moisture. Large temperature changes, when associated with high levels of humidity, can lead to the deposition of moisture on components such as circuit boards, with disastrous results.

This situation can easily arise in hot climates: the plant cools down overnight and dew is deposited on cold surfaces. Some protection may be afforded by continuous air conditioning, the use of silica gel or alumina driers[11]. These granular substances are strongly hygroscopic and are capable of achieving very low relative humidities. However, their capacity for absorbing moisture is of course limited and the container should be removed regularly for regeneration by a hot air stream.

Legionnaires' Disease

This disease is a severe form of pneumonia and infection is usually the result of inhaling water droplets carrying the causative bacteria *(Legionella pneumophila)*. Factors favouring the organism in water systems are the presence of deposits such as rust, algae and sludge, a temperature between 20 and 45°C and the presence of light. Clearly all these conditions can be present in systems involving cooling towers.

Preventive measures include:

- treatment of water with scale and corrosion inhibitors to prevent the build up of possible nutrients for the organisms
- use of suitable water disinfectant such as chlorine 1–2 ppm or ozone
- steam humidifiers are preferable to water spray units.

If infection is known to be present flushing, cleaning and hyper-chlorination are necessary. If a system has been out of use for some time, heating to about 70–75°C for 1 h will destroy any organisms present.

This matter should be taken seriously. This is one of the few cases in which the operators of a test facility may be held guilty of endangering life, with consequent ruinous claims for compensation.

Summary

Design of the ventilation system for a test cell is a major undertaking and a careful and thorough analysis of expected heat loads from engine, exhaust system, dynamometer, instrumentation, cooling fans and lights is essential if subsequent difficulties are to be avoided.

A separate air supply for the engine should always be considered.

Air conditioning is expensive but may be necessary, both on the grounds of comfort and to regulate the state of the air entering the engine.

The ventilation of the control room should ensure that conditions there meet normal office standards.

Notation

atmospheric pressure	p_a bar
atmospheric temperature	$t_a\,°C$
density of air	$\rho\,kg/m^3$
gas constant for air	$R = 287\,J/kgK$
specific heat of air at constant pressure	$C_p\,kJ/kgK$
mass rate of flow of air	$m\,kg/s$
rate of heat loss, vertical surface	$Q_v\,W/m^2$
rate of heat loss, horizontal surface	$Q_h\,W/m^2$
temperature of surface	$t_s\,°C$
rate of heat loss by radiation	$Q_r\,W/m^2$
ventilation air temperature rise	$\Delta T°C$
total heat load	$H_L\,kW$
ventilation air flow rate	$Q_A\,m^3/s$
fan air flow rate	$Q_F\,m^3/s$
velocity of air	$V\,m/s$
velocity pressure	$p_v\,Pa$
static pressure	$p_s\,Pa$
total pressure	$p_t\,Pa$
pressure loss	$\Delta p_t\,Pa$
static air power	$P_{s_F}\,kW$
total air power	$P_{t_F}\,kW$
shaft power	$P_A\,kW$
moisture content	$\omega\,kg/kg$
relative humidity	ϕ
specific enthalpy of moist air	$h\,kJ/kg$
cooling load	$L_C\,kW$
heating load	$L_H\,kW$
emissivity	ε
pressure loss coefficient	K_e

References

1. Pita, E.G. (1981) *Air Conditioning Principles and Systems: an Energy Approach*, Wiley, Chichester.

2. Heinsohn, R.J. (1991) *Industrial Ventilation Engineering Principles*, Wiley, Chichester.
3. Bejan, A. (1993) *Heat Transfer*, Wiley, Chichester.
4. Freeston, H.G. (1958) Test bed installations and engine test equipment, *Proc. I. Mech.E.*, **172** (7).
5. TM 8 *Design Notes for Ductwork*, Chartered Institution of Building Services, London.
6. *Industrial Ventilation: a Manual of Recommended Practice* (1982) (17th edn) American Conference of Government Industrial Hygienists: Committee on Industrial Ventilation.
7. BS 5720 *Code of Practice for Mechanical Ventilation and Air Conditioning in Buildings*.
8. BS 848 Part 1 *Fans for General Purposes*.
9. *C.I.B.S. Psychrometric Chart*, Chartered Institution of Building Services, London.
10. BS 7120 *Specification for Rating and Performance of Air to Liquid and Liquid to Liquid Chilling Packages*.
11. BS 2540 *Specification for Granular Desiccant Silica Gel*.

Further reading

BS 599 *Methods of Testing Pumps*.
BS 6339 *Specification for Dimensions of Circular Flanges for General Purpose Industrial Fans*.

5   Vibration and noise

Vibration is considered with particular reference to the design of engine mountings and the isolation of engine-induced disturbances.

The theory of noise generation and control is briefly considered, and a brief account given of the particular problems involved in the design of anechoic cells.

Vibration and noise

Almost always the engine itself is the only significant source of vibration and noise in the test cell[1-5]. Secondary sources such as the ventilation system, pumps and circulation systems or the dynamometer are usually swamped by the effects of the engine. There are several aspects to this problem:

- The engine must be mounted in such a way that neither it nor connections to it can be damaged by excessive movement or excessive constraint.
- Transmission of engine-induced vibration to the cell structure or to other buildings must be controlled.
- Excessive noise levels in the cell should be avoided as far as possible and the design of alarm signals should take noise levels into account.

Fundamentals: sources of vibration

Since the vast majority of engines likely to be encountered are single- or multi-cylinder in-line vertical engines we shall concentrate on this configuration. An engine may be regarded as having six degrees of freedom of vibration about orthogonal axes through its centre of gravity: linear vibrations along each axis and rotations about each axis, see Fig 5.1. In practice only three of these modes are usually of importance:

- vertical oscillations on the X axis due to unbalanced vertical forces
- rotation about the Y axis due to cyclic variations in torque
- rotation about the Z axis due to unbalanced vertical forces in different transverse planes

Torque variations will be considered later. In general the rotating masses are

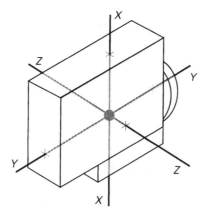

Figure 5.1. *Internal combustion engine: principal axes and degrees of freedom*

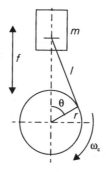

Figure 5.2. *Connecting rod crank mechanism: unbalanced forces*

carefully balanced but periodic forces due to the reciprocating masses cannot be avoided. A crank, connecting rod and piston assembly, Fig. 5.2, is subject to a periodic force in the line of action of the piston given approximately by:

$$f = m_p\omega_c^2 r \cos\theta + \frac{m_p\omega_c^2 r \cos 2\theta}{n} \qquad \text{where } n = l/r \qquad (1)$$

Here m_p represents the sum of the mass of the piston plus, by convention, one third of the mass of the connecting rod (the remaining two thirds is usually regarded as being concentrated at the crankpin centre).

The first term of eq. (1) represents the first order inertia force. It is equivalent to the component of centrifugal force on the line of action generated by a mass m_p concentrated at the crankpin and rotating at engine speed. The

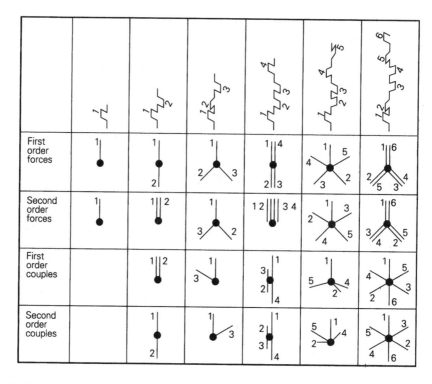

Figure 5.3. *First and second order forces, multi-cylinder engines*

second term arises from the obliquity of the connecting rod and is equivalent to the component of force in the line of action generated by a mass $m/4n$ at the crankpin radius but rotating at twice engine speed.

Inertia forces of higher order ($3\times, 4\times$, etc. crankshaft speed) are also generated but may usually be ignored.

It is possible to balance any desired proportion of the first order inertia force by balance weights on the crankshaft but these then give rise to an equivalent reciprocating force on the Z axis, which may be even more objectionable.

Inertia forces may be represented by vectors rotating at crankshaft speed and twice crankshaft speed. Figure 5.3 shows the first and second order vectors for engines having from one to six cylinders. Note the following features:

- In a single cylinder engine both first and second order forces are unbalanced.
- For larger numbers of cylinders first order forces are balanced.
- for two and four cylinder engines the second order forces are unbalanced and additive.

This last feature is an undesirable characteristic of a four cylinder engine, and in some cases has been eliminated by counter-rotating weights driven at twice crankshaft speed.

The other consequence of reciprocating unbalance is the generation of rocking couples about the transverse or Z axis and these are also shown in Fig. 5.3:

- There are no couples in a single cylinder engine.
- In a two cylinder engine there is a first order couple.
- In a three cylinder engine there are first and second order couples.
- Four and six cylinder engines are fully balanced.
- In a five cylinder engine there is a small first order and a larger second order couple.

Six cylinder engines, which are well known for smooth running, are balanced in all modes.

Variations in engine turning moment are discussed in Chapter 8. These variations give rise to equal and opposite reactions on the engine, which tend to cause rotation of the whole engine about the crankshaft axis. The order of these disturbances, i.e. the ratio of the frequency of the disturbance to the engine speed, is a function of the engine cycle and the number of cylinders. For a four-stroke engine the lowest order is equal to half the number of cylinders: in a single cylinder there is a disturbing couple at half engine speed while in a six cylinder engine the lowest disturbing frequency is at three times engine speed. In a two-stroke engine the lowest order is equal to the number of cylinders.

The design of engine mountings

The main problem in engine mounting design is that of ensuring that the motions of the engine and the forces transmitted to the surroundings as a result of the unavoidable forces and couples briefly described above are kept to manageable levels. In the case of vehicle engines it is sometimes the practice to make use of the same flexible mounts and the same location points as in the vehicle; this does not, however, guarantee a satisfactory solution. In the vehicle the mountings are carried on a comparatively light structure, while in the test cell they may be attached to a massive pallet or even to a seismic block. Also in the test cell the engine may be fitted with additional equipment and various service connections. All of these factors alter the dynamics of the system when compared with the situation of the engine in service and can give rise to fatigue failures.

Truck diesel engines usually present less of a problem than small automotive engines, as they generally have fairly massive and well-spaced supports at the flywheel end. Stationary engines will in most cases be carried on four or more

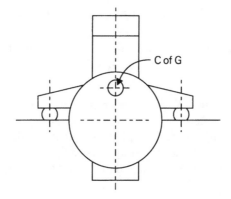

Figure 5.4.   *Internal combustion engine carried on four flexible mountings*

flexible mountings in a single plane below the engine and the design of a suitable system is a comparatively simple matter.

We shall consider the simplest case, an engine of mass m kg carried on undamped mountings of combined stiffness k N/m, Fig. 5.4. The differential equation defining the motion of the mass equates the force exerted by the mounting springs with the acceleration of the mass:

$$\frac{md^2x}{dt^2} + kx = 0 \tag{1}$$

a solution is

$$x = \text{constant} \times \sin\sqrt{\frac{k}{m}} \cdot t$$

$$\frac{k}{m} = \omega_0^2 \qquad \text{natural frequency} = n_0 = \frac{\omega_0}{2\pi} = \frac{1}{2\pi}\sqrt{\frac{k}{m}} \tag{2}$$

the static deflection under the force of gravity $= mg/k$ which leads to a very convenient expression for the natural frequency of vibration:

$$n_0 = \frac{1}{2\pi}\sqrt{\frac{g}{\text{static deflection}}} \tag{3}$$

or, if static deflection is in millimetres:

$$n_0 = \frac{15.76}{\sqrt{\text{static deflection}}} \tag{3a}$$

This relationship is plotted in Fig. 5.5.

Next consider the case where the mass m is subjected to an exciting force of amplitude f and frequency $\omega/2\pi$. The equation of motion now reads:

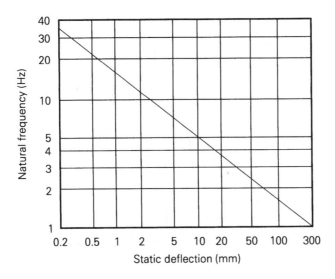

Figure 5.5. *Relation between static deflection and natural frequency*

$$m\frac{\mathrm{d}^2 x}{\mathrm{d}t^2} + kx = f \sin \omega t$$

the solution includes a transient element; for the steady state condition amplitude of oscillation is given by:

$$x = \frac{f/k}{(1 - \omega^2/\omega_0^2)} \tag{4}$$

here f/k is the static deflection of the mountings under an applied load f. This expression is plotted in Fig. 5.6 in terms of the amplitude ratio, $x \div$ static deflection. It has the well-known feature that the amplitude becomes theoretically infinite at resonance, $\omega = \omega_0$.

If the mountings combine springs with an element of viscous damping, the equation of motion becomes:

$$m\frac{\mathrm{d}^2 x}{\mathrm{d}t^2} + c\frac{\mathrm{d}x}{\mathrm{d}t} + kx = f \sin \omega t$$

where c is a damping coefficient. The steady state solution is:

$$x = \frac{f/k}{\sqrt{\left(1 - \dfrac{\omega^2}{\omega_0^2}\right)^2 + \dfrac{\omega^2 c^2}{mk\omega_0^2}}} \sin(\omega t - A) \tag{5}$$

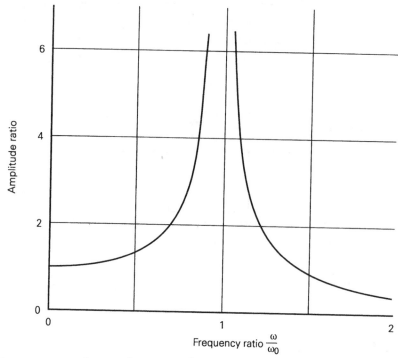

Figure 5.6. *Relation between frequency and amplitude ratio (trans-missibility), undamped vibration*

If we define a dimensionless damping ratio:

$$C^2 = \frac{c^2}{4mk}$$

this equation may be written:

$$x = \frac{f/k}{\sqrt{\left(1 - \frac{\omega^2}{\omega_0^2}\right)^2 + 4C^2\frac{\omega^2}{\omega_0^2}}} \sin(\omega t - A) \qquad (5a)$$

(if $C = 1$ we have the condition of critical damping when, if the mass is displaced and released it will return eventually to its original position without over-shoot).

The amplitude of the oscillation is given by the first part of this expression:

$$\text{amplitude} = \frac{f/k}{\sqrt{\left(1 - \frac{\omega^2}{\omega_0^2}\right)^2 + 4C^2\frac{\omega^2}{\omega_0^2}}}$$

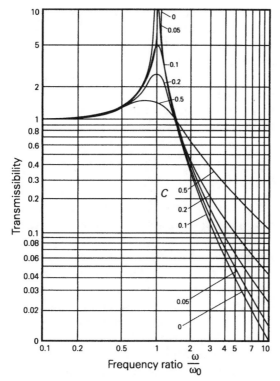

Figure 5.7. *Relation between transmissibility (amplitude ratio) and frequency, damped oscillations for different values of damping ratio C (logarithmic plot)*

This is plotted in Fig. 5.7 together with the curve for the undamped condition, Fig. 5.6 and various values of C are shown. The phase angle A is a measure of the angle by which the motion of the mass lags or leads the exciting force. It is given by the expression:

$$A = \tan^{-1} \dfrac{2C}{\dfrac{\omega_0}{\omega} - \dfrac{\omega}{\omega_0}} \tag{6}$$

At very low frequencies A is zero and the mass moves in phase with the exciting force. With increasing frequency the motion of the mass lags by an increasing angle, reaching 90° at resonance. At high frequencies the sign of A changes and the mass leads the exciting force by an increasing angle, approaching 180° at high ratios of ω to ω_0.

Natural rubber flexible mountings have an element of internal (hysteresis) damping which corresponds approximately to a degree of viscous damping corresponding to $C = 0.05$.

The essential role of damping will be clear from Fig. 5.7: it limits the potentially damaging amplitude of vibration at resonance. The ordinate in Fig. 5.7 is often described as the *transmissibility* of the mounting system: it is a measure of the extent to which the disturbing force f is reduced by the action of the flexible mounts. It is considered good practice to design the system so that the minimum speed at which the machine is to run is not less than three times the natural frequency, corresponding to a transmissibility of about 0.15. It should be noticed that once the frequency ratio exceeds about 2 the presence of damping actually has an adverse effect on the isolation of disturbing forces.

Practical considerations in the design of engine mountings

In the above simple treatment we have only considered oscillations in the vertical direction. In practice, as has already been pointed out, an engine carried on flexible mountings has six degrees of freedom, Fig. 5.1. While in many cases a simple analysis of vibrations in the vertical direction will give a satisfactory result, under test cell conditions a more complete computer analysis of the various modes of vibration and the coupling between them may be advisable. This is particularly the case with tall engines with mounting points at a low level, when cyclic variations in torque may induce transverse rolling of the engine.

Reference (6) lists the design factors to be considered in planning a system for the isolation and control of vibration and transmitted noise:

- specification of force isolation
  - as attenuation, dB
  - as transmissibility
  - as efficiency
  - as noise level in adjacent rooms
- natural frequency range to achieve the level of isolation required
- load distribution of the machine
  - is it equal on each mounting?
  - is the centre of gravity low enough for stability?
  - exposure to forces arising from connecting services, exhaust system, etc.
- vibration amplitudes–low frequency
  - normal operation
  - fault conditions
  - starting and stopping
  - is a seismic block or sub-base needed?
- higher-frequency structure-borne noise (100Hz+)
  - is there a specification?
  - details of building structure
  - sufficient data on engine and associated plant
- transient forces

- shocks, earthquakes, machine failures
- environment
  - temperature
  - humidity
  - fuel and oil spills.

Detailed design of engine mountings for test bed installation is a highly specialized matter, see Ker Wilson[1] for guidance on standard practice. In general the aim is to avoid 'coupled' vibrations, e.g. the generation of pitching forces due to unbalanced forces in the vertical direction, or the generation of rolling moments due to the torque reaction forces exerted by the engine. These can give rise to resonances at much higher frequencies than the simple frequency of vertical oscillation calculated in the following section and to consequent trouble, particularly with the engine-to-brake connecting shaft.

Massive foundations

Heavy concrete foundations carried on flexible mountings (seismic blocks) are very expensive to install, calling for deep excavations, elaborate shuttering and elaborate arrangements, such as tee-slotted bases, for bolting down the engines. With the wide range of different types of flexible mounting now available, it is questionable whether, except in special circumstances, such as a requirement to install test facilities in close proximity to offices, their use is economically justified. The trough surrounding the block may be of incidental use for installing services.

The analysis and prediction of the extent of transmitted vibration to the surroundings is a highly specialized field. The theory is dealt with in ref.1, the starting point being the observation that a heavy block embedded in the earth has a natural frequency of vibration that generally lies within the range 1000 to 2000 c.p.m. There is thus a possibility of vibration being transmitted to the surroundings if exciting forces, generally associated with the reciprocating masses in the engine, lie within this frequency range. An example would be a four cylinder four-stroke engine running at 750 rev/min: we see from Fig. 5.3 that such an engine generates substantial second order forces at twice engine speed, or 1500 c.p.m. Fig 5.8, redrawn from ref. 1, gives an indication of acceptable levels of transmitted vibration from the point of view of physical comfort.

Fig. 5.9 is a sketch of a typical seismic block. Reinforced concrete weighs roughly 2500 kg/m^3 and this block would weigh about 4500 kg. Note that the surrounding tread plates must be isolated from the block, also that it is essential to earth the mounting rails. The block is shown carried on four combined steel spring and rubber isolators, each having a stiffness of 100 kg/mm, Fig. 5.10. From equation (2a) the natural frequency of vertical oscillation of the bare block would be 4.70 Hz, or 282 c.p.m., so the block would be a

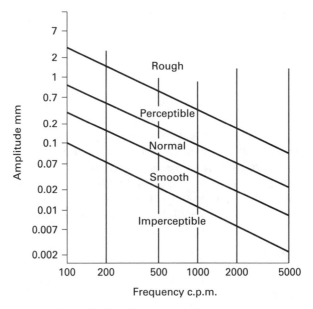

Figure 5.8. *Perception of vibration*

suitable base for an engine running at about 900 rev/min or faster. If the engine weight were, say, 500 kg the natural frequency of block + engine would be reduced to 4.46 Hz, a negligible change.

Fig. 5.11 shows a rather similar construction. In this case the block is supported on natural cork pads having a surface area of $1 \, m^2$ and a thickness of 50 mm. The compression of the pads will be about 0.5 mm, giving a natural frequency in the region of 20 Hz. Another possibility is the use of rubber sheeting for the same purpose.

A special application concerns the use of seismic blocks for supporting engines in anechoic cells. It is in principle good practice to mount engines undergoing noise testing as rigidly as possible, since this reduces noise radiated as the result of movement of the whole engine on its mountings, and the lowering of the centre of gravity is similarly helpful. Fig. 5.12 shows a lead filled anechoic frame sometimes used for this purpose.

While the dynamometer is not a significant source of vibration it is sometimes the practice to mount both engine and brake on a common block.

Finally it should be remarked that there is available on the market a bewildering array of different designs of isolator or flexible mounting, based on steel springs, natural or synthetic rubber of widely differing properties used in compression or shear, and combinations of these materials. For the non-specialist the manufacturer's advice should be sought.

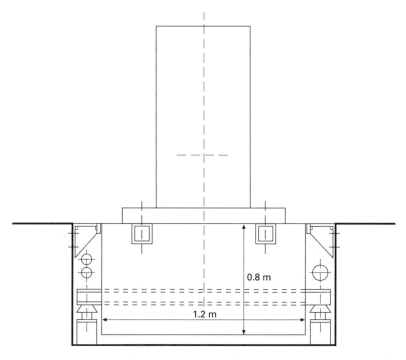

Figure 5.9. *Spring-mounted seismic block*

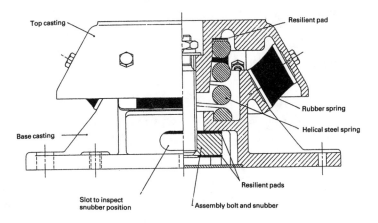

Figure 5.10. *Combined spring and rubber flexible mounting*

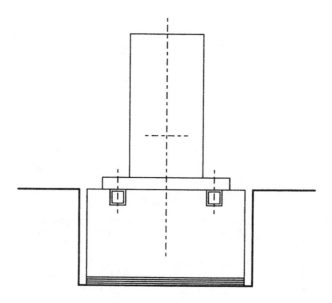

Figure 5.11.   *Seismic block on cork pad*

Summary

This section should be read in conjunction with Chapter 8, which deals with the associated problem of torsional vibrations of engine and dynamometer. The two aspects – torsional vibration and other vibrations of the engine on its mountings – cannot be considered in isolation.

The exciting forces arising from (inevitable) unbalance in a reciprocating engine are considered and a description of the essential features of mounting design is given, together with a check list of points to be considered.

Noise: fundamentals[3,4]

Sound intensity

The starting point in the definition of the various quantitative aspects of noise measurement is the concept of sound intensity, defined as:

$$I = \frac{\bar{p}^2}{\rho c} \text{ W/m}^2$$

where p^2 is the mean square value of the acoustic pressure, i.e. the pressure variation due to the sound wave, ρ the density of air and c, the velocity of sound in air.

① Mounting plate   ⑤ Sub frame pallet
② Anti-vibration pads   ⑥ Bearing block assembly
③ Inertia base, inc. lead weights   ⑦ Shaft guard
④ Sub frame

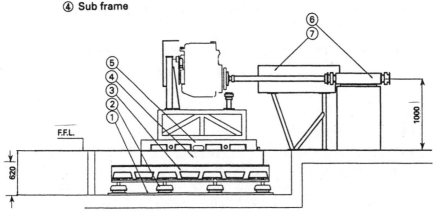

Figure 5.12. *Inertia base assembly for anechoic cell.*

Intensity is measured in a scale of *decibels* (dB):

$$\mathrm{dB} = 10 \log_{10} \left(\frac{I}{I_0} \right) = 20 \log_{10} \left(\frac{\bar{p}}{p_0} \right)$$

where I_0 corresponds to the average lower threshold of audibility, taken by convention as $I_0 = 10^{-12}$ W/m², an extremely low rate of energy propagation.

From these definitions it is easily shown that a doubling of the sound intensity corresponds to an increase of about 3 dB ($\log_{10} 2 = 0.301$). A tenfold increase gives an increase of 10 dB while an increase of 30 dB corresponds to a factor of 1000 in sound intensity. It will be apparent that I varies through an enormous range.

The value on the decibel scale is often referred to as the *sound pressure level* (SPL). In general sound is propagated spherically from its source and the inverse square law applies. Doubling the distance results in a reduction in SPL of about 6 dB ($\log_{10} 4 = 0.602$).

The human ear is sensitive to frequencies in the range from roughly 16 Hz to 20 kHz, but the perceived level of a sound depends heavily on its frequency structure. The well-known *Fletcher–Munson curves,* Fig. 5.13, were obtained by averaging the performance of a large number of subjects who were asked to decide when the apparent loudness of a pure tone was the same as that of a reference tone of frequency 1 kHz. Loudness is measured in a scale of *phons,* which is only identical with the decibel scale at the reference frequency. The decline in the sensitivity of the ear is greatest at low frequencies. Thus at 50 Hz an SPL of nearly 60 dB is needed to create a sensation of loudness of 30 phons.

Acoustic data are usually specified in frequency bands one octave wide. The standard mid-band frequencies are:

31.5   62.5   125   250   500   1000   2000   4000   8000   16 000 Hz

e.g. the second octave spans 44–88 Hz. The two outer octaves are rarely used in noise analysis.

Noise measurements

Most instruments for measuring sound contain weighted networks which give a response to frequency which approximates to the Fletcher–Munson curves. In other words their response to frequency is a reciprocal of the Fletcher–Munson relationship, Fig. 5.13. For most applications the *A*-weighting curve gives satisfactory results and the corresponding SPL readings are given in *dBA*. *B*- and *C*-weightings are sometimes used for high sound levels while a special *D*-weighting is used primarily for aircraft noise measurements, Fig. 5.14.

The dBA value gives a general 'feel' for the intensity and discomfort level of a noise but for analytical work the unweighted results should be used. The simplest type of sound level meter for diagnostic work is the *octave band analyser*. This instrument can provide flat or *A*-weighted indications of SPL for each octave in the standard range. For more detailed study of noise emissions an instrument capable of analysis in one-third octave bands is more effective. With such an instrument it may for example be possible to pinpoint a particular pair of gears as a noise source.

For serious development work on engines, transmissions or vehicle bodies much more detailed analysis of noise emisssions is provided by the discrete Fourier transform (DFT) or fast Fourier transform (FFT) digital spectrum analyser. The mathematics on which the operation of these instruments is based is somewhat complex but fortunately they may be used effectively without a detailed understanding of the theory involved. It is well known that any periodic function, such as the cyclic variation of torque in an internal combustion engine, may be resolved into a fundamental frequency and a series of harmonics (see Chapter 8). General noise from an engine or transmission does not repeat in this way and it is accordingly necessary to record a sample of the noise over a finite interval of time and to process this data to give a spectrum of SPL against frequency. The Fourier transform algorithm allows this to be done.

Permitted levels of noise

Noise levels actually within an engine test cell nearly always exceed the levels

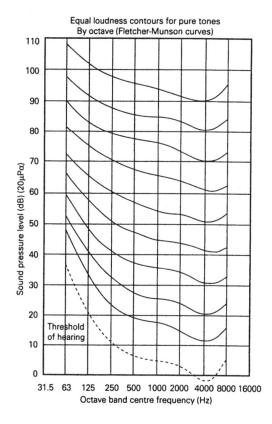

Figure 5.13. *Fletcher–Munson curves of equal loudness*

permitted by statute[7], while the control room noise level must be kept under observation and appropriate measures taken. UK regulations require the provision of noise protection at noise levels between 85 and 90 dBA while above these levels the use of hearing protection is mandatory. European and US regulations specify ways of assessing the mean sound level over a working day and give permitted values for this mean. Warring[5] quotes the following decibel levels corresponding to various noisy situations:

Concorde take-off at 500 m	120 dB
Disco, car horn at 1 m	110 dB
Heavy industry	100 dB
Underground train	90 dB
Heavy city traffic	70 dB

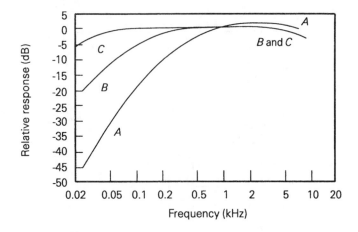

Figure 5.14. *Noise weighting curves*

Table 5.1. *Absorption coefficients at an octave band centre frequency 1 kHz*

Concrete, brickwork	0.03
Glass	0.1
Breeze blocks	0.5
Acoustic tiles	0.9
Open window	1.0

Noise in the test cell environment

The measured value of SPL in an environment such as an engine test cell gives no information as to the power of the source: a noisy machine in a cell having good sound-absorbent surfaces may generate the same SPL as a much quieter machine surrounded by sound-reflective walls. The *absorption coefficient* is a measure of the sound power absorbed when a sound impinges once upon a surface. It is quite strongly dependent on frequency and tends to fall as frequency falls below about 500 Hz. Information on absorption coefficients for a wide range of structural materials and sound insulators is given in IHVE Guide B 12[8]. A few approximate values are given in Table 5.1 and indicate the highly reverberatory properties of untreated brick and concrete.

It should be remembered that the degree to which sound is absorbed by its surroundings makes no difference to the intensity of the sound received directly from the engine.

'Cross talk' between test cell and control room or other adjacent rooms can occur through any openings in the partition walls and also through air conditioning ducts when there is a common system.

Vehicle noise measurements

A number of regulations lay down permitted vehicle noise levels and the methods by which they are to be measured[9,10]. In the EC it is specified that *A*-weighted SPL should be measured during vehicle passage at a distance of 7.5 m from the centreline of the vehicle path. There are also regulations for noise measurement near the exhaust pipe outlet and in the driving compartment.

Permitted noise levels for passing vehicles range from 77 dBA for passenger cars up to 84 dBA for large trucks.

Special test cells for noise testing

The testing of engines and vehicles for *noise, vibration and harshness* (NVH) forms a substantial element of vehicle development programmes, and calls for specialized (and very expensive) test facilities. Essentially such a cell should provide an environment approximating as closely as possible to that of the vehicle on the road.

Such cells are of two kinds: *semi-anechoic*, (in the USA. commonly *hemi-anechoic*) in which walls and ceiling are lined with sound-absorbent materials while the floor is reflective, generally of concrete, and *full anechoic cells*, in which all surfaces including the floor are sound absorbent. (A full anechoic cell is adapted to measuring the noise radiated in every direction from a source, while the semi-anechoic cell simulates the situation where the source is located in the open but resting on a reflective horizontal plane; clearly the latter is appropriate for land vehicle testing).

The design of anechoic cells is a matter for the specialist, but BS 4196[11], while specifically concerned with methods of determining sound power levels of noise sources, gives some useful guidance on certain aspects of anechoic cell design, including:

desirable shape and volume of the cell,
desirable absorption coefficient of surfaces,
specification of absorptive treatment,
guidance regarding avoidance of unwanted sound reflections.

There are certain other points that should be borne in mind by the non-specialist if he is required to take responsibility for setting up a test facility of this kind:

- The structure of an anechoic cell should be isolated from any environment in which noise is generated such as production plant. Common ventilation systems should be avoided.

- The internal volume of the building shell will be roughly twice that of the usable space because of the space required for the acoustic lining. This lining often takes the form of foam shapes which may protrude some 700 mm from the supporting structure, which may itself be some 300 mm from the cell wall.

- Access doors and windows affect the acoustic performance adversely and must be kept to a minimum commensurate with safety. The acoustic lining covering exits should be coloured to make their location obvious. Windows should preferably be avoided since glass has poor sound-absorbent properties, see Table 5.1.

- Acoustic research makes use of highly sensitive instruments and correspondingly sensitive measurement channels. The design of the signal cabling must be carefully considered at the design stage to avoid interference and a multitude of trailing leads in the cell.

- The ventilation of acoustic cells presents particular problems. In most ordinary test work the noise contributed by the ventilation system may be ignored but this is not the case with acoustic cells, where the best solution is usually to dispense with a forced draught fan and to draw in the air by way of suitable filters and silencers, using only an induced draught fan. However, full power running in anechoic cells is rarely prolonged and the precise control of inlet air temperature is not so critical as in cells concerned with power measurement.

- Acoustic cell linings usually absorb liquids and may be inflammable, increasing fire hazards. Particular attention should be paid to smoke detection and to the fire quenching system.

- The dynamometer must clearly be isolated acoustically from the cell proper. For work on engines and transmissions the brake is located outside the cell, calling for a long coupling shaft, the design of which can present problems. For work on complete vehicles a rolling road dynamometer, Chapter 16, will be the usual solution: this will transmit a certain amount of noise to the cell which may be measured by motoring the rolls in the absence of a vehicle.

- The engine in an anechoic cell must be raised above floor level, typically by 1.0–1.2 m to shaft centreline, to permit microphones to be located below it. This calls for non-standard engine mounting arrangements.

Exhaust noise

The noise from test cell exhaust systems can travel considerable distances and be the subject of complaints from neighbouring premises, particularly if running takes place at night or during weekends. The design of test cell exhaust systems is largely dictated by the requirement that the performance of engines under test should not be adversely affected, see p.112, and the additional requirement that noise emission should be a minimum may be hard to meet.

Essentially there are two types of device for reducing the noise level in ducts: resonators and absorption mufflers. A resonator, sometimes known as a reactive muffler, is shown in Fig. 5.15. It consists of a cylindrical vessel divided by partitions into two or more compartments. The exhaust gas travels through the resonator by way of perforated pipes which themselves help to dissipate noise. The device is designed to give a degree of attenuation, which may reach 50 dB, over a range of frequencies, see Fig. 5.16 as a typical example.

Absorption mufflers consist essentially of a chamber lined with sound-absorbent material through which the exhaust gases are passed in a perforated pipe. Absorption mufflers give broad-band damping but are less effective than resonators in the low-frequency range. However, they offer less resistance to flow.

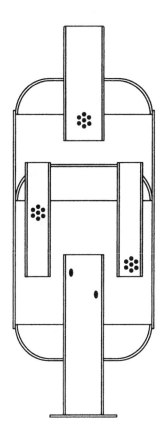

Figure 5.15. *Exhaust resonator or reactive muffler*

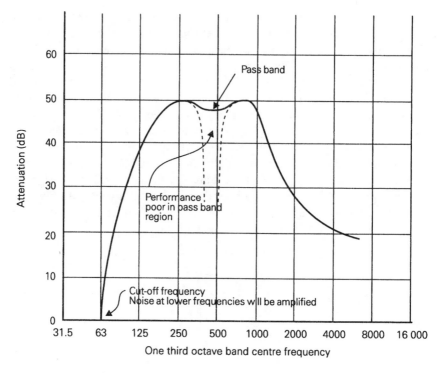

Figure 5.16. *Performance of reactive muffler*

Selection of the most suitable designs for a given situation is a matter for the specialist. Both types of silencer are subject to corrosion if not run at a temperature above the dew point of the exhaust gas and condensation in an absorption muffler is particularly to be avoided.

Summary

This brief statement of the basic principles of noise measurement and theory will serve as an introduction to the non-specialist. Applications include noise control in buildings, vehicles and exhaust systems.

Notation

Vibration

mass of piston + 1/3 connecting rod m_p kg
angular velocity of crankshaft ω_c s^{-1}

crank radius	r m
crank angle from tdc	θ
connecting rod length $\div$ r	n
unbalanced exciting force	f N
mass of engine	m kg
combined stiffness of mountings	k N/m
amplitude of vibration	x m
angular velocity of vibration	ω rad/s
angular velocity at resonance	ω_0 rad/s
natural frequency	n_0 Hz
phase angle	A rad
damping coefficient	c N s/m
damping ratio	C
acceleration due to gravity	g m/s^2

Noise

r.m.s. value of acoustic pressure	$\bar{p}$ N/m^2
density of air	ρ kg/m^3
velocity of sound in air	c m/s
sound intensity	I W/m^2
threshhold sound intensity	I_0 W/m^2

References

1. Ker-Wilson, W. (1959) *Vibration Engineering,* Griffin, London.
2. Thompson, W.T. (1988) *Theory of Vibration,* 3rd edn, Prentice-Hall, London.
3. Fader, B. (1981) *Industrial Noise Control,* Wiley, Chichester.
4. Turner, J.D. and Pretlove, A.J. (1991) *Acoustics for Engineers,* Macmillan Education, London.
5. Warring, R.H. (1983) *Handbook of Noise and Vibration Control,* 5th edn, Trade and Technical Press, Morden, Surrey.
6. Maw, A.N. (1992) The design of resilient mounting systems to control machinery noise in buildings, *Plastics, Rubber and Composites Processing and Applications,* **18**, 9–16.
7. *Noise at Work Regulations* (1990) Health and Safety Executive, London.
8. I.H.V.E. Guide B12: *Sound Control,* Chartered Institution of Building Services, London.
9. BS 3425 *Method for the Measurement of Noise emitted by Motor Vehicles.*
10. BS 3539 *Specification for Sound Level Meters for the Measurement of Noise Emitted by Motor Vehicles.*
11. BS 4196 Parts 0 to 8 *Sound Power Levels of Noise Sources.*

Further reading

BS 2475 *Specification for Octave and One-third Octave Band-pass Filters.*
BS 3045 *Method of Expression of Physical and Subjective Magnitudes of Sound or Noise in Air.*
BS 4198 *Method for Calculating Loudness.*
BS 4675 Parts 1 and 2 *Mechanical Vibration in Rotating Machinery.*
BS 5330 *Method of Testing for Estimating the Risk of Hearing Handicap due to Noise Exposure.*

6 Flow systems: fuel, water, combustion air and exhaust

This chapter describes the essential features of a test installation fuel supply system and mentions the many statutory regulations governing their design. The thermodynamics of water cooling are considered and the calculation of cooling water requirements is described. A note on water quality follows, and the design of test cell cooling water systems is outlined. Recommendations are made regarding the supply of combustion air and the design of exhaust systems is described. Finally, to pull together the various topics considered in the first six chapters of the book an energy balance is drawn up for a complete test cell and the process of sizing the various services is described.

Fuel supply systems

This is an area subject to extensive statutory regulations, see for example *Petrol Filling Stations; Construction and Operation* published by the Health and Safety Executive[1,2]. Any engineer taking responsibility for anything beyond the smallest and simplest installation should make himself familiar with all relevant regulations of this kind, both national and local.

Both the legislative requirements and the way in which they are interpreted by local officials can vary widely, even in a single country. Where engine test installations are well established there should be a good understanding of the requirements whereas fire and planning officers who have no experience of the industry can react with concern and may require tactful guidance.

Design requirements for fuel oil storage tanks are laid down in BS 799[3] which lists a variety of tanks of different shapes and sizes. Figure 6.1 shows a typical arrangement for a fuel oil or gasolene storage tank in accordance with this Standard. Horizontal cylindrical mild steel welded tanks more than 1 m in diameter may alternatively be fabricated to BS 2594[4]. All tanks should have their internal surfaces treated prior to installation to avoid contamination with rust or sand blast material. Storage tanks for LPG should conform to BS 5500[5].

For the storage of fuel in drums or other containers a petroleum store such as is shown in Figure 6.2 is to be recommended. This meets the requirements of the Licensing Department of the former Greater London Council. The lower part forms a well, not more than 0.6 m deep, capable of containing the total

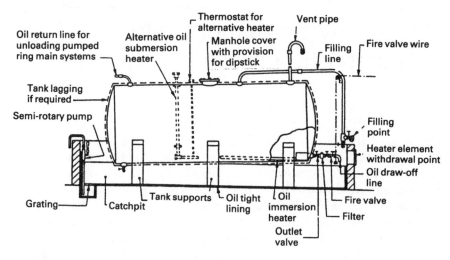

Figure 6.1. *Typical fuel storage tank*

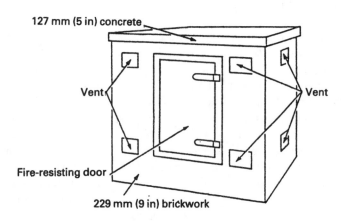

Figure 6.2. *Recommended design of fuel store*

volume of fuel authorized to be kept in the store. Ventilators at high and low level are to be covered by fine wire gauze mesh and protective grilles.

The use of standard (non galvanized) drawn steel tubing for fuel lines that remain full is entirely satisfactory, but if they are likely to spend appreciable periods partially drained the use of stainless steel should be considered. The use of threaded fittings is not to be recommended, although with care and the use of modern sealants they can be satisfactory provided there is no significant pipe movement, particularly thermally induced movement. The use of any kind of fibrous 'pipe jointing' should be absolutely forbidden as fibre contamination is difficult to clear.

Preferably, all fuel lines, and certainly all underground lines, should be constructed with welded joints and flanges, or with the use of compression fittings approved for the fuels concerned (some fuels such as 'winter diesel' appear to be particularly penetrating). External fuel lines should be lagged and must be trace heated if temperatures are likely to fall to a level at which fuel 'waxing' may take place.

Underground fuel lines

Buried fuel lines should be wrapped with water-repellent bandage (U.K., 'Denzo Tape') and laid in fine gravel in a well compacted trench. However good modern practice is to run fuel lines in a sealed trench of concrete sections with a load-bearing lid.

There are certain special requirements for the fuel supply system for an engine test facility that may not apply to systems for other purposes. Provision must be made for the use of a number of different fuels, and the typical 'fuel farm' will be provided with a number of storage tanks and several ring mains. Much use is made of special and reference fuels supplied by the drum and it is convenient to draw directly from these.

The fuel system in the test cell will vary widely in complexity. For the occasional test a simple fuel tank of the type used for outboard engines, capacity not more than 10 litres, may be all that is necessary. In some cases the system may be limited to a single fuel line connected to the fuel injection pump but a special purpose test cell may call for the supply of many different fuels, transient fuel consumption measurement and fuel temperature control.

Fuel supplies to a cell may be provided either under static head from a day tank or by a pressurized fuel system fed from a central fuel farm. Day tanks must be fuel-specific to prevent cross-contamination so there may be a requirement for several. Modern safety practice dictates that day tanks should be kept outside the test facility and be fitted with both a dump valve for operation in case of fire and a monitored overflow and return system.

The minimum static head is commonly about 4.5 m. Since the tanks are exposed to ambient weather conditions they should be shielded from direct sunlight and, in the case of diesel fuels, lagged or trace-heated. If a pumped system is used it must be remembered that for much of the operating life the fuel demand will be below the rated flow and the system must be able to operate under bypass conditions without cavitation or undue heating.

In designing systems to meet this wide range of requirements some general principles should be borne in mind:

- each fuel line penetrating the cell wall should be provided with a normally closed solenoid operated valve interlocked with both the cell control system and the fire protection circuits
- it may be necessary to provide several separate fuel supplies to a cell; a typical provision would be three lines for diesel fuel, leaded and unleaded gasolene respectively. Problems can arise from carry-over, particularly when changing from leaded to unleaded fuel, with consequent danger of 'poisoning' exhaust catalysers. The capacity of the system, including such items as filters, needs to be kept in mind. To minimize cross contamination it is desirable to locate the common connection as close to the engine as possible. Each line will have a wall-mounted sub-system and the isolating valve, mentioned above, may also be used for fuel selection
- it is good practice to have a cumulative fuel meter in each line for general control and for contract charging
- air entrainment and vapour locking can be a problem in test cell fuel systems. An air eliminating valve should be fitted at the highest point in the system, with an unrestricted vent to atmosphere external to the cell
- if control of fuel temperature is required the same comments as those made concerning engine coolant apply, p.107, but the materials of all units such as heat exchangers, and of sealing materials, must be checked with the manufacturer as to their suitability for use with fuel. It is particularly important to minimize the distance between the temperature controlling element and the engine so that, if running is interrupted, the engine receives fuel at the desired temperature with the minimum of delay
- a simple fuel heating system may comprise a small header tank with electric heater to provide hot water at controlled temperature for circulation through the water-to-fuel heat exchanger. Such an arrangement gives good temperature control despite wide variations in fuel flow rate
- breaking of fuel lines, with consequent spillage, should as far as possible be avoided. It is sensible to mount all components on a permanent wall mounted panel, with switching via interlocked and clearly marked valves. The run of the final flexible fuel lines to the engine should not interfere with operator access. A common arrangement is to run the lines from the dynamometer end of the stand; alternatively an overhead boom may be used. Self-sealing couplings should be used for engine and other frequently broken connections
- it is essential to fit oil drain traps to waste connections to avoid the possibility of discharging oil or fuel into the foul water drains
- maximum recommended fuel line velocity: 0.2 m/s.

Fuel measuring devices are discussed in Chapter 9.

Gaseous fuels

Liquefied petroleum gas (LPG)

Liquefied petroleum gas consists mainly of propane. Its vapour pressure at 40°C is about 15 bar and storage vessels are designed for a working pressure of about 20 bar. Detailed instructions for the storage and distribution of LPG are given in refs 6 and 7 and should be followed with care.

The main LPG hazards are fire and explosion. The flash point is below −40°C and explosivity limits are approximately 1.9 to 9.5% by volume in air.

The liquid has a marked tendency to build up a static charge when transferred by pipelines and it is essential to earth pipelines, receiving and transfer vessels.

Small fires in LPG systems should be dealt with by dry powder extinguishers, but it is also essential to cool storage vessels exposed to flame.

Natural gas, compressed natural gas (NC, CNG)

Natural gas consists of about 90% methane, which has a boiling point at atmospheric pressure of −163°C, and LNG ships are designed to transport the gas refrigerated to approximately this temperature. Engine test installations requiring natural gas usually draw this from a mains supply at just above atmospheric pressure so high-pressure storage arrangements are not necessary. Fire hazards are moderate when compared with LPG installations. As a safety precaution with gaseous fuels the use of 'sniffer' type sensors may be recommended.

Residual fuels

Many large stationary diesel engines and the majority of marine engines operate on heavy residual fuels that require special treatment before use and must also be heated before delivery to the fuel injection system. In such cases the test cell fuel supply system is required to incorporate the special features of the fuel supply and treatment systems installed in diesel propelled ships.

It is always necessary with fuels of this type to remove sludge and water before use in the engine. The problem here is that the density of residual fuels can approach that of water, making separation very difficult. The accepted procedure is to raise the temperature of the oil, thus reducing its density, and then to feed it through a purifier and clarifier in series. These are centrifugal devices, the first of which removes most of the water while the second completes the cleaning process.

References 1 and 8 give detailed recommmendations for the design and operation of such systems and Fig. 6.3 taken from ref. 8 shows a schematic arrangement.

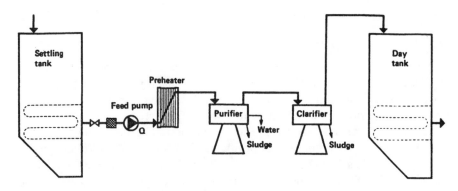

Figure 6.3.   *Fuel supply system for heavy fuel oil*

It is also necessary to provide change-over arrangements so that the engine can be started and shut down on a light fuel.

Water supply systems

Fundamentals

Water is the ideal liquid cooling medium. Its specific heat is higher than that of any other liquid, roughly twice that of hydrocarbons. It is of low viscosity, relatively non-corrosive and of course readily available.

The specific heat of water is usually taken as:

$$C = 4.1868 \text{ kJ/kg K}$$

(this is in fact the value of the 'international steam table calorie' and corresponds to the specific heat at 14°C. The specific heat is very slightly higher at each end of the liquid range: 4.21 kJ/kg K at 0°C and at 95°C but these variations may be neglected).

The use of antifreeze (ethylene glycol) permits operation over a wide range of coolant temperatures. A 50% by volume solution of ethylene glycol in water permits operation down to a temperature of −33°C. Antifreeze also raises the boiling point of the coolant and a 50% solution will operate at a temperature of 135°C with pressurization of only 1.5 bar.

The specific heat of ethylene glycol is about 2.28 kJ/kg K and, since its density is 1.128 kg/l, the specific heat of a 50% by volume solution is:

$$0.5 \times 4.1868 + 0.5 \times 2.28 \times 1.128 = 3.38 \text{ kJ/kg K}$$

or 80% of that of water alone. Thus the circulation rate must be increased by 25% for the same heat transfer rate and temperature rise.

The relation between flow rate, q_w(litres/hour), temperature rise, Δt, and heat transferred to the water is:

$$4.1868 q_w \Delta t = 3600\, H$$
$$q_w \Delta t = 860\, H \tag{1}$$

where H = heat transfer rate, kW
(To absorb $1\,kW$ with a temperature rise of $10°C$ the required flow rate is thus $86\,l/h$).

Required flow rates

It is good practice to limit the temperature rise of the cooling water through the engine water jacket to about $10°C$. In the case of the dynamometer the flow rate is determined by the maximum permissible cooling water temperature, since it is desirable to avoid the deposition of scale (temporary hardness) on the internal surfaces of the machine. Eddy current dynamometers, in which the heat to be removed is transferred through the walls, are more sensitive in this respect than hydraulic machines, in which heat is generated within the cooling water.

Recommended maximum (leaving) temperatures are:

Eddy current machines $60°C$
Hydraulic dynamometers $70°C^{*}$

(provided carbonate hardness of water does not exceed $50\,mg$ CaO/l, 1.25 British Units. For greater hardness values limit temperatures to $50°C$).

Approximate cooling loads per kilowatt of engine power output (see p.210) are shown in Table 6.1.

Corresponding flow rates and temperature rises are as shown in Table 6.2.

Water quality[9]

At an early stage in planning a new test facility it is essential to ensure that a sufficient supply of water of appropriate quality can be made available. Control of water quality, which includes the suppression of bacterial infections, algae and slime, is a complex matter and it is advisable to consult a water treatment expert who is aware of local conditions. If the available water is not of suitable quality then the programme must include the provision of water treatment plant.

Most dynamometer manufacturers publish tables, prepared by a water chemist, which specify the water quality required for their machines. The following notes are intended for the guidance of non-specialists in the subject.

---

* At this temperature some machines can experience internal flash boiling.

Table 6.1.   *Cooling loads*

	Output (kW/kW)
Automotive gasolene engine, water jacket	0.9
Automotive diesel engine, water jacket	0.7
Medium speed heavy diesel engine	0.4
Oil cooler	0.1
Hydraulic or eddy current dynamometer	0.95

Table 6.2.   *Cooling water flow rate*

	In °C	*Out °C*	*l/kWh*
Automotive gasolene engine	70	80	75
Automotive diesel engine	70	80	60
Medium speed heavy diesel engine	70	80	35
Oil cooler	70	80	5
Hydraulic dynamometer	20	68	20
Eddy current dynamometer	20	60	20

Solids in water

Circulating water should be as free as possible from solid impurities. If water is to be taken from a river or other natural source it should be strained and filtered before entering the system. Raw surface water usually has significant turbidity caused by minute clay or silt particles which are ionized and may only be removed by specialized treatments (coagulation and flocculation). Other sources of impurities include drainage of dirty water into the sump, windblown sand entering cooling towers and casting sand from engine water jackets. Hydraulic dynamometers are sensitive to abrasive particles and accepted figures for the permissible level of suspended solids are in the range 2 to 5 mg/litre. The use of seawater or estuarine water is not to be recommended other than in specially designed marine installations.

Water hardness

The hardness of water is a complex property, not easy to measure objectively. Hard water, if its temperature exceeds about 70°C, may deposit calcium carbide 'scale' (temporary hardness), which can be very destructive to all types of dynamometer and heat exchanger. A scale deposit greatly interferes with heat transfer and commonly breaks off into the water flow when it can jam control valves and block passages. Soft water may have characteristics that cause corrosion, so very soft water is not ideal either.

Essentially, hardness is due to the presence of divalent cations, usually calcium or magnesium, in the water. When a sample of water contains more than 120 mg of these ions per litre, expressed in terms of calcium carbonate, $CaCO_3$, it is generally classified as a hard water.

There are several national scales for expressing hardness, but at present no internationally agreed scale:

American and British
$1°$ US $= 1°$ UK $= 1$ mg $CaCO_3$ per kg water $= 1$ ppm $CaCO_3$

French
$1°F = 10$ mg $CaCO_3$ per litre water

German
$1°G = 10$ mg CaO per litre of water

(the old British system, 1 Clarke degree $= 1$ grain per Imperial gallon $= 14.25$ ppm $CaCO_3$)

Requirements for dynamometers are usually specified as within the range 2 to 5 Clarke degrees (30 to 70 ppm $CaCO_3$)

Water may be either acid or alkaline/basic. Water molecules, HOH (commonly written H_2O), have the ability to dissociate, or ionize, very slightly. In a perfectly neutral water (neither acidic nor basic) equal concentrations of H+ and OH− are present. The pH value is a measure of the hydrogen ion concentration: its value is important in almost all phases of water treatment, including biological treatments. Acid water has a pH value of less than 7.07 and most dynamometer makers call for a pH value in the range 7 to 9; the ideal is within the range 8 to 8.4.

The preparation of a full specification of the chemical and biological properties of a given water supply is a complex matter. Many compounds – phosphates, sulphates, sodium chloride and carbonic anhydride – all contribute to the nature of the water, the anhydrides in particular being a source of dissolved oxygen that may make it aggressively corrosive. This can lead to such problems as the severe roughening of the loss plate passages in eddy current dynamometers, causing failures due to local water starvation and plate distortion (the narrow passages in eddy current dynamometers are particularly liable to blockage which can arise from chemicals used in some water treatment regimes).

Control of water quality also includes the suppression of bacterial infections, algae and slime (Legionnaire's disease, p.67).

BS 4959[10], describes the additives used to prevent corrosion and scale formation, with chemical tests for the control of their concentration and also gives guidance on the maintenance and cleaning of cooling water systems.

A recirculating system should include a small bleed-off to drain, to prevent deterioration of the water by concentration of undesirable elements. A bleed rate of about 1% of system capacity per day should be adequate. If no bleed-off is included the entire system should be periodically drained, cleaned out and refilled with fresh water.

Design of test cell cooling water circuits

Cooling water circuits may be classifed as follows, wiith increasing levels of complexity:

- service modules and cooling columns involving some loss of mains water
- unpressurized engine and dynamometer cooling water circuits
  (in the above arrangements the engine water jacket is cooled by the same water as circulates in the dynamometer)
- open circuits incorporating service modules with heat exchangers for engine water jacket (pressurized) cooling systems and for lubricant cooling
- pressurized multiple circuits using closed cooling towers or heat exchangers for use with eddy current dynamometers or a.c./d.c. water cooled machines.

Cooling columns and service modules

If special engine coolants are not required a cooling column, Fig. 6.4, is a simple and economical solution. It can be portable and located close to the engine under test. The column allows the engine outlet temperature to run up to its designed level, when a thermostatic valve opens, allowing cold water to enter the bottom of the column and hot water to run to waste from the top. The top of the column is fitted with a standard automotive radiator cap for use when filling the engine circuit.

Unpressurized or 'open' cooling water circuits

Fig. 6.5 shows a typical system. It is based on a sump lying below floor level and divided into hot and cold areas by a partition wall. There should be a small opening in the bottom of the wall for drain-down purposes and the top should lie about 50 mm below normal working level. A rough rule for deciding sump capacity is that the water should not be turned over more than once per minute. Sufficient excess capacity above working level should be provided to accommodate drain-back from engines and dynamometers.

A pump draws water from the hot well by way of a strainer and delivers it to an external evaporative cooler. The cooled water is returned to the cold well. Water for engine and dynamometer cooling is drawn from the cold well and drains back into the hot well. There is a continuous loss of water due to evaporation plus the small drainage to waste mentioned above and makeup is supplied by way of a float valve.

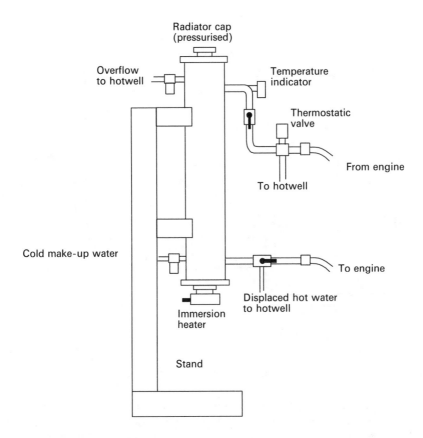

Figure 6.4. *Cooling column*

To minimize air entrainment the pump suction should be located close to a corner, return flow should be by way of a submerged pipe with air vent, and it is an advantage to subdivide the sump by perforated screens between water return and suction sections. Finally, consideration should be given at the design stage to the consequences of a power failure. Consider, for example, the consequences of a sudden failure in the water supply to a hydraulic dynamometer absorbing 10 MW at full speed from a marine diesel engine. The system will take some time to bring to rest, during which the brake will be operating on a mixture of air and water vapour, with the possibility of serious overheating. Wherever possible there should be some provision for gravity feed in the event of a sudden power failure.

It is desirable that the supply pressure to hydraulic dynamometers should be stable or the control of the machine will be affected. This implies that the supply pump must be of adequate capacity, with a flat pressure-volume

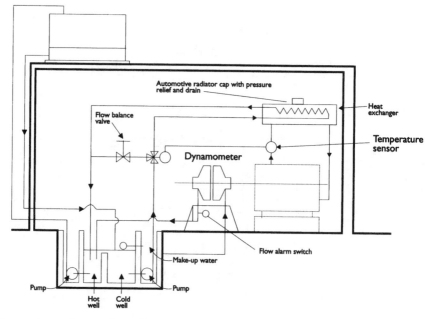

Figure 6.5.   *Test cell cooling water circuit*

characteristic. Sufficient pressure must be available under maximum demand conditions to meet the specified dynamometer head.

The design and installation of a water supply for a large test installation is a specialist task not to be underestimated. It requires the specification of a large number of test and flow balancing valves, together with stand-by pumps and filters with changeover arrangements.

Open circuits plus service modules with heat exchanger

Most test cell cooling water circuits isolate the engine jacket cooling water circuit from the general cooling water system. This has the following advantages:

● special coolants may be used in the engine circuit
● with correct design and sizing of the secondary cooling system much more precise control of cooling water temperature may be achieved than is possible with an open system
● secondary circuits for lubricating oil and fuel temperature control, etc., may be run at any desired temperature.

Fig. 6.6(a) is an illustration of a typical service module incorporating heat exchangers for jacket coolant and lubricating oil, while Fig. 6.6(b) shows a simplified schematic of the circuit. The combined header tank and heat

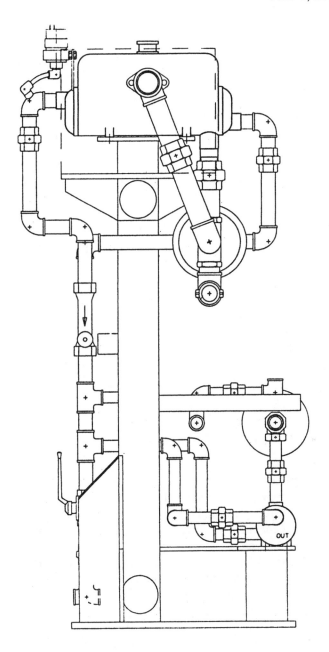

Figure 6.6(a). *Typical service module*

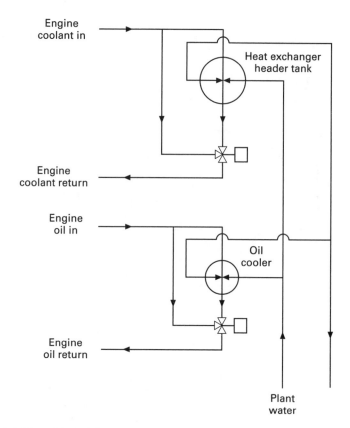

Figure 6.6(b). *Circuit diagram*

exchanger is a particularly useful feature. This has a filler cap and relief valve and acts in every way as the equivalent of a conventional engine radiator and ensures that the correct pressure is maintained. If some engines are to be tested without their own coolant pumps the module must be fitted with a circulating pump, commonly of the type used in central heating systems. For ease of maintenance, it should be possible to withdraw exchanger tube stacks without major dismantling of the system, and simple means for draining both oil and coolant circuits should be provided.

Certain special requirements apply to oil cooling units. The entire circuit must lie below engine sump level so that there is no risk of flooding the sump and it may be necessary to provide heaters in the circuit for rapid warm-up of the engine. In this case precautions must be taken to ensure that the oil cannot in any circumstances be overheated with danger of 'cracking'. Fig. 6.7 shows schematically a separate lubricating oil cooling and conditioning unit.

While it may in cases where very accurate temperature control is necessary

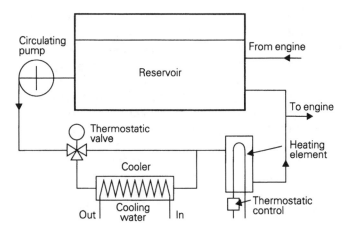

Figure 6.7. *Lubricating oil cooling and conditioning circuit*

be appropriate to use separate pallet mounted cooling modules located close to the engine, permanently located oil and cooling water modules offer the best compromise for most engine testing. One common position is behind the dynamometer where both units may be fed from the external cooling water system and the engine connection hoses may be run under the dynamometer to a connection point near the shaft end.

Pressurized multiple cooling water circuits

While all hydraulic dynamometers must operate with a free discharge of cooling water, eddy current dynamometers can be used with a closed pressurized cooling system. Design of such systems is a matter for specialist building services engineers. The specification must ensure a constant flow to the dynamometers despite variations in thermal load. The cooling water must be treated to avoid 'gumming up' of the internal channels of the machine.

Engine coolant and oil temperature control

Precise control of coolant temperature is not easily achieved except in the special case in which the service module is designed to match the thermal characteristics of the engine with which it is associated; even here it is difficult to achieve stable temperatures at light load. The instability is increased if the engine is much smaller than that for which the cooling circuit is designed. The capacity of the heat exchanger is the governing factor and it may be advisable, when a wide range of engine powers is to be accommodated, to provide several coolers with a range of capacities.

The most usual arrangement is to control the temperature by means of a three-way thermostatically controlled valve. The alternative, where temperature is controlled by regulating the primary cooling water flow, has the advantage of not disturbing the engine coolant flow rate but gives an inherently lower rate of response to load changes. Time and distance lag, between a sensor located at the engine outlet and the control valve at the cooler, may be significant and the pipe runs between engine and service module should be kept to a minimum. In cases where temperature control is critical for the particular test sequence it may be necessary to arrange a pallet-mounted cooling module close to the engine. More usually, cooling modules are mounted behind the dynamometer with a common feed and drain, and the engine connections are run beneath the dynamometer. In cases such as anechoic cells, where the heat exchanger is inevitably remote from the engine, it may be necessary to speed up the rate of circulation by an auxiliary pump to reduce lag.

A proportional and integral (P and I) controller will give satisfactory results in most cases but it is important that thermostatic valves should be correctly sized, since they will not function satisfactorily at flow rates much lower than that for which they are designed. Ensure that the valve is installed with flow in the correct direction.

The sum of these phenomena is often referred to as the 'thermal inertia' of the cooling system. Compared with most common industrial process control systems the engine test cell cooling system is 'fast': it is easier to highlight the problems of temperature control than to supply answers that will work in every case.

Flow velocities in cooling water systems

The maximum water velocity in a supply system should not exceed 3 m/s to avoid the possibility of cavitation, but should reach a minimum of 1.5 m/s to sweep away deposited matter. Velocities in gravity drains should not exceed 0.6 m/s. It is good practice to keep pipe runs as straight as possible and to use bends rather than elbows.

Design of water-to-water heat exchangers

Manufacturers of heat exchangers invariably provide simple design procedures for establishing sizes and flow rates for a given heat exchanger performance. If it should for any reason be desirable to design a heat exchanger from first principles ref. 11 gives a detailed design procedure and a useful discussion of factors affecting cost.

Health hazards associated with cooling towers

The serious lung infection known as Legionnaires' disease results from inhalation of droplets of infected water from cooling towers or evaporative condensers. See p.67

Combustion air

The influence of the condition of the combustion air (its pressure, temperature, humidity and purity) on engine performance is discussed in Chapter 10, where it is shown that variations in these factors can have a very substantial effect on performance. In an ideal world engines under test would all be supplied with air at 'standard' conditions, p.184. In practice there is a trade-off between the advantages of such standardization and the cost of achieving it.

For routine production testing variations in the condition of the air supply are not particularly important, but the performance recorded on the test document should be corrected to standard conditions, see p.187. For research and development testing, however, it is desirable that, so far as is practicable, the combustion air should be supplied at constant conditions of temperature and humidity (atmospheric pressure remains in practice beyond our control).

The simplest and most widely used method of supplying the combustion air is to allow the engine to take its supply directly from the test cell atmosphere. The great advantage of using cell air for vehicle engines in particular is that rigging of complex 'in vehicle' air filtration and ducting units is straightforward.

The major disadvantage of drawing the air from within the cell is the uncontrolled variability in temperature and quality arising from air currents and other disturbances in the cell. These can include contamination with exhaust and other fumes and may be aggravated by the use of spot fans. A secondary disadvantage is the imposition of a variable load on the cell ventilation system which can, if this system is not under direct pressure and temperature control, lead to significant variations in cell environment.

These disadvantages are commonly overcome by the provision of a separate, possibly temperature conditioned, supply drawn from the external atmosphere and terminating in a cone end which is arranged to envelop the air inlet. The amount of air to be supplied to such a system depends to some extent on the type of test sequence. If a fixed volume system is designed to deliver twice the maximum engine demand the supply will be effectively 'flooded' but there may be prolonged periods of running with a considerable excess supply and associated waste of energy. Also this excess air must be taken into account in the total cell energy calculations, Chapter 4.

A close coupled air supply duct may be necessary in certain cases, for example in anechoic cells where air intake noise must be eliminated, or where a precisely controlled conditioned supply is needed, but its design presents problems. The duct must be of sufficient size to avoid an appreciable pressure drop during engine acceleration and conversely there should be arrangements to spill excess air externally to the cell on a sudden reduction in demand.

This last point is important: engine air consumption varies more or less directly with speed, a parameter that can vary more rapidly than any air supply system can respond, and it is essential that this system should not impose pressure changes on the air inlet. Pressure pulsations in the inlet ducting can also give rise to serious disturbances, see p.194.

In any system designed to bring conditioned air into the cell consideration must be given to the danger of condensation forming in the system during shut down periods. Underfloor ducts, and those in which there is any possibility of condensation draining into the engine, should if possible be avoided.

Conditioning of combustion air

In much R and D work on small and medium sized engines it is not acceptable either to ignore atmospheric conditions or to monitor these conditions and apply correction factors. Examples are tests in which very small differences/improvements in performance are being investigated and tests in which humidity of the combustion air is a major factor, as may be the case in emissions testing and work involving the study of corrosion or condensation.

Air conditioning systems are bulky and are commonly roof mounted externally. A typical system will be designed to control the combustion air temperature within the range 20 to 25°C with an accuracy of ±1 deg C. It will contain the following main components:

> air inlet with screen and bellmouth
> heater, either electrical or hot water
> cooling element, with chilled water supply
> fan unit

Systems for the supply of combustion air of which both temperature and humidity is controlled are expensive, the expense increasing with the range of conditions to be covered and the degree of precision required. They are also energy intensive, particularly when it is necessary to reduce the humidity of the atmospheric air by cooling and condensation. Before introducing such a system a careful analysis should be made to ensure that it is really justified.

To indicate more exactly what is involved in full conditioning of combustion air, let us suppose that it is necessary to supply air at standardized conditions to the 250 kW diesel engine for which the energy balance is given in Table 11.5, p.211. The calculated air consumption of the engine is 1312 kg/hr, corresponding to a volumetric flow rate of about 0.3 m^3/s. The necessary components of an air conditioning duct for attachment to the engine air inlet are shown diagrammatically in Fig. 6.8 and comprise in succession:

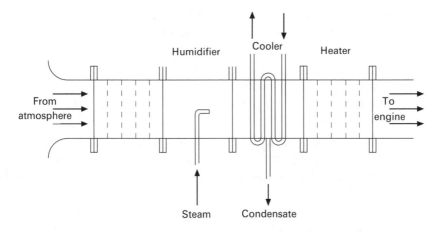

Figure 6.8.   *Air conditioning duct for conditioning air*

air inlet with screen and bellmouth
heater, either electrical or hot water
humidifier, comprising steam injector
with associated steam generator
cooling element, with chilled water supply
secondary heating element

It may be asked why it is necessary to include two heater elements, one before
and one after the humidifier. The first element is necessary when it is required
to humidify cold dry air, since if steam is injected into cold air supersaturation
will result and moisture will immediately be deposited. On the other hand it is
commonly necessary, in order to dry moist air to the desired degree, to cool it
to a temperature lower than the desired final temperature and reheat must be
supplied downstream of the cooler.

The internal dimensions of the duct in the present example would be
approximately $0.3\,\text{m} \times 0.3\,\text{m}$. Assume that the air is to be supplied to the
engine at the standard conditions specified in BS 5514 (ISO 3046)[12]:

temperature 25°C
relative humidity (r.h.) 30%

Let us consider two fairly extreme atmospheric conditions and determine the
conditioning processes necessary:

hot and humid: 35°C and 80% r.h.
cool to 7°C, cooling load 34 kW
(flow of condensate 0.5 l/min)
reheat to 25°C, heating load 6.5 kW

it is necessary to cool to 7°C in order to reduce the moisture content, originally 0.030 kg water/kg air, to the required value, 0.006 kg water/kg air; the latter corresponds to saturation at 7°C.

cold and dry: 0°C and 50% r.h.
heat to 25°C, heating load 9 kW
inject steam at rate of 0.1 l/min,
heating load 4 kW

It will be observed that the energy requirements, particularly for the chiller, are quite substantial. In the case of large engines and gas turbines any kind of combustion air conditioning is not really practicable and reliance must be placed on correction factors.

Centralized combustion air supplies

Centralized combustion air conditioning units can be designed to supply air at 'standard' conditions to a number of cells, and this is often a cost-effective solution. If a range of conditions is to be provided affecting individual cells, however, individual conditioning units are necessary, although they may use common sources for chilled water and steam.

Health and safety: fire control

Any combustion air supply system must be integrated with the fire alarm and fire extinguishing system, p.22, and must include the same provisions for isolating individual cells as the main ventilation system.

Exhaust systems

It is possible to run into operational and safety problems with test cell exhaust systems and there have been accidents, some of them fatal. From the research engineer's viewpoint, the ideal exhaust system for an engine under test is one that resembles exactly the system that would be used on the same engine in service.

Many modern automotive engines are fitted with emission control systems which require that the actual vehicle exhaust system should be employed: this can impose difficult problems even in cells of large floor area. Such a system is shown schematically in Fig. 6.9(b). The practical problems of locating the scavenge duct in a position that can be reached by the vehicle tail pipe can be considerable and often call for modification of the vehicle system. However some major engine manufacturers impose stringent rules as to permissible modifications, for example insisting that the number and angle of any bends should not be changed and the overall length not altered.

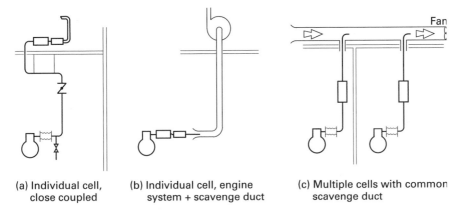

(a) Individual cell,   (b) Individual cell, engine   (c) Multiple cells with common
    close coupled   system + scavenge duct   scavenge duct

Figure 6.9. *Exhaust systems*

If vehicle systems are to be modified it should be remembered that any change in the length of the primary pipe is particularly undesirable, since this can lead to changes in the exhaust and induction processes as a result of changes in the pattern of exhaust pulses in the system. This can affect the volumetric efficiency and power output of the engine and, in the case of two-stroke engines, it may prove impossible to run the engine at all with a wrongly proportioned exhaust system.

An increase in the length of pipe beyond the first chamber is less critical but care should be taken to limit the back-pressure imposed upon the engine, as measured by a manometer at the silencer inlet, to a maximum of perhaps 100 mm H_2O; any greater restriction will probably have a perceptible effect on volumetric efficiency and power output.

It is often difficult to choose a convenient run for the exhaust system in the cell. Turbocharged engines in particular have complex exhaust systems and run at such high temperatures that large areas of manifold and exhaust pipe can appear incandescent and this can represent a large heat load on the ventilating system. The system should be guarded to avoid burns to staff and vulnerable equipment. The use of bandage type lagging may be appropriate providing it does not cause overheating of exhaust system components.

Unless the system has been specifically designed with this in view, it is not generally a good idea to run the exhaust pipe in a floor duct (and extremely dangerous if the duct also carries fuel lines).

The practice sometimes adopted, of discharging the engine exhaust into the main ventilation extraction duct has, in the opinion of the authors, several disadvantages:

- the duct must be of stainless steel, to avoid rapid corrosion
- the air flow must be increased to a level greater than that necessary for basic

ventilation, to maintain an acceptable duct temperature. On a cold day this can lead to chilling in the cell

- soot deposits in the fan can cause problems
- other difficulties can arise, such as noise, variability in exhaust back pressure, etc.

If 'tail pipe' testers of the type used in vehicle servicing and inspection are to be employed then easy access to the tail pipe end is necessary.

Test cell exhaust systems may be classified as follows:

- individual cell, close coupled
- individual cell, engine system + scavenged duct
- multiple cells with common scavenged duct.

Individual cell, close coupled

Such an arrangement is shown schematically in Fig. 6.9(a). It may be regarded as the 'standard' arrangement for a general purpose testbed, and is also suitable for production testing. The exhaust manifold is coupled to a flexible stainless steel pipe, of fairly large diameter to minimize pressure waves, and led by way of a back-pressure regulating valve to a pipe system suspended from the cell roof. Condensate, which is highly corrosive, tends to collect in these pipes, which should be laid to a fall with suitable drainage arrangements.

It is a requirement of most exhaust emissions instrumentation that the sample is taken from the exhaust pipe at the correct point; some devices have quite specific requirements regarding size, position and angle of probe insertion and it is desirable, in order to ensure a representative sample, that there should be 6 diameters length of straight pipe both upstream and downstream of the probe. It is good practice when designing the exhaust system to arrange a straight horizontal run of exhaust pipe above head height, this pipe to be easily replaced by a pipe with specific probe tappings. Most exhaust analysis equipment also permits only a limited length of (heated) pipeline between the sampling point and the analysis unit. This limitation may require the cabinet containing the gas analysis equipment to be located within the cell.

Individual cell, engine system + scavenged duct

When it is considered necessary to use the vehicle exhaust system two options are available: one is to take the pipe outside the building through a panel in the cell wall, the other to use a scavenge air system as shown schematically in Fig. 6.9(b). In this case the tailpipe is simply inserted into a bell mouth through which cell air is drawn; the flow rate should be at least twice the maximum exhaust flow, preferably more. This outflow should be included in the calculations of cell ventilation air flow. The scavenge flow is induced by a fan, usually centrifugal, which must be capable of handling the combined air and exhaust flow at temperatures that may reach 150°C

Multiple cells + common scavenged duct

This arrangement, Fig 6.9(c), may be regarded as standard for large installations. To illustrate the design process for the scavenging system, let us suppose we are setting up an installation of three test cells each like that illustrated in Fig 6.10, p. 116. These are to be used for the production testing of the 250 kW turbocharged diesel engine for which maximum exhaust flow rate is shown in the energy balance calculation, p. 211, as 1365 kg/hour. To ensure adequate dilution and a sufficiently low temperature in all circumstances, we need to cater for the possibility of all three engines running at full power simultaneously and a scavenge air flow rate of about 10 000 kg/hr, say 2.3 m³/sec would be appropriate. Table 4.2, p. 53, indicates a flow velocity in the range 15–20 m/s and hence a duct size in the region of 400 mm × 300 mm, or 400 mm diameter. As before, the scavenging fan must be suitable for temperatures of at least 150°C.

This arrangement is recommended only for diesel engines. In the case of spark ignition engines there is always the possibility that unburned fuel, say from an engine that is being motored, could accumulate in the ducting and then be ignited by the exhaust from another engine. The possibility may seem remote, but accidents of this kind are by no means unknown.

Note particularly that, in cases (b) and (c), the fan controls must be interlocked with the cell control systems so that engines can only be run when the fan is running.

Where the engine's own exhaust system is not used the section of exhaust tubing adjacent to the engine must be flexible enough to allow the engine to move on its mountings and a stainless steel bellows section is to be recommended. Exhaust tubing used in this area should be regarded as expendable and the workshop should be equipped to make up replacements and fit transducers.

As a final point, silencers that are much oversized for the capacity of the engine will never get really hot and will be rapidly destroyed by corrosion.

The increasing demand for vehicle engines to be tested with the exhaust system used in the vehicle can present hazards, with hot exhaust pipes laid at near floor level, and suitable guarding should be available.

Drawing up the energy balance and sizing the system

In Chapter 1 it is recommended that at an early stage in the design of a new test cell a diagram similar to Figs. 1.2 or 1.3. should be drawn up to show all the flows: air, fuel, water, exhaust gas, electricity and heat into and out of the cell.

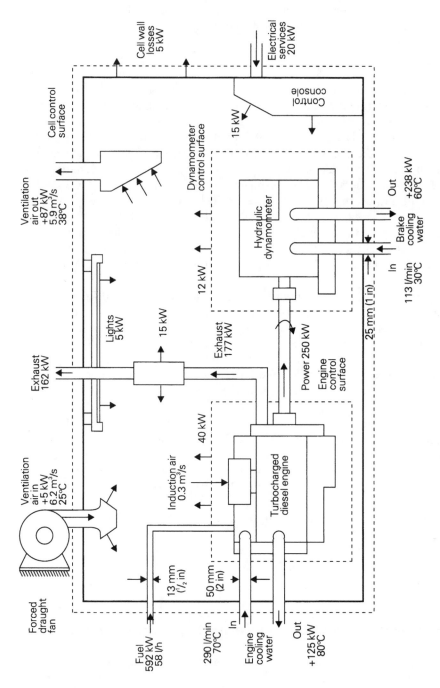

Figure 6.10. *Energy balance and flow diagram for 250 kW test cell*

The process is best illustrated by an example, and the full-load regime for a 250 kW turbocharged diesel engine driving a hydraulic dynamometer has been chosen. The heat balance for the engine itself is derived in Chapter 11 and shown diagrammatically in Fig. 11.7, p.212. This shows the results of the engine cooling water calculation, the exhaust energy, the induction air flow and the estimated convection and radiation losses.

The design of the ventilation system for the cell is dealt with in Chapter 4, and the corresponding air and heat flows are shown in Fig. 6.8. The cooling water flow for the hydraulic dynamometer may be calculated on the basis of the following assumptions:

95% of power absorbed appears in the cooling water
Cooling water inlet temperature 30°C
Maximum desirable outlet temperature 70°C

Then flow to brake:

$$\frac{250 \times 0.95 \times 60}{4.186 \times 40} = 85 \, 1/\text{min}$$

The electrical power input to the cell, for lights, instrumentation and spot fans is assumed to be 20 kW and, finally, the heat losses through the cell walls and ceiling are assumed to be small, 5 kW.

Calculated pipe sizes are as follows, see p.108:

Fuel, velocity 0.2 m/s, 10.8 mm dia, say 1/2 in
Engine cooling water, 3 m/s, 45 mm dia, say 2 in
Brake cooling water, 3 m/s, 28 mm dia, say 1 in

A final point regarding engine cooling water: it may be assumed that this will make use of a service module, Figs 6.4 or 6.5. Either way the overall energy flow will be unchanged but in the former case the temperature rise of the primary cooling water may be substantially higher, leading to a smaller flow rate.

A final warning regarding sizing. The above example is sized for an engine running at full power. At light loads the control characteristics of, for example, the engine cooling water and oil cooler systems may be unsatisfactory. If full power will very rarely be used it may be preferable to design for a rather lower load and tolerate limited ability to run at full load, or higher than desired temperatures.

Summary

The design of various services concerned with fluids, fuel and water, and also air to the engine and exhaust leaving it, is outlined. Precautions regarding fuel and economy measures regarding water are discussed. The possible dangers associated with exhaust systems are emphasized and finally an example of cooling water system design and sizing is given.

References

1. Hughes, J.R. and Swindells, S. (1987) *The Storage and Handling of Petroleum Liquids,* Griffin, London.
2. *Petrol Filling Stations: Construction and Operation,* Health and Safety Executive, London.
3. BS 799 Part 5 *Specification for Oil Storage Tanks.*
4. BS 2594 *Specification for Carbon Steel Welded Horizontal Cylindrical Storage Tanks.*
5. BS 5500 *Specification for Unfired Fusion Welded Pressure Vessels.*
6. Standard No. 58: *Storage and Handling of Liquefied Petroleum Gas,* National Fire Protection Agency, London.
7. *Liquefied Petroleum Gas. Product Dossier No. 92/102,* Institute of Petroleum, London.
8. *Recommendations for Cleaning and Pretreatment of Heavy Fuel Oil,* Alfa Laval Ltd, London.
9. Vesilind, P.A. *et al.* (1988) *Environmental Engineering,* Butterworth-Heinemann, Oxford.
10. BS 4959 *Recommendations for Corrosion and Scale Prevention in Engine Cooling Water Systems.*
11. McAdams, W.H. (1973) *Heat Transmission,* McGraw-Hill, Maidenhead.
12. BS 5514 *Reciprocating Internal Combustion Engines: Performance.*

Further reading

Bejan, A. (1993) *Heat Transfer,* Wiley, Chichester.

7 Choosing the right dynamometer

Perhaps the most difficult question facing the engineer setting up a test facility is the choice of the most suitable dynamometer. In this chapter the characteristics, advantages and disadvantages of the various types are discussed and a procedure for arriving at the correct choice is described.

The earliest form of dynamometer, the rope brake, Fig. 7.1(a), dates back to the early years of the last century. An extremely dangerous device, it was nevertheless capable of giving quite accurate measurements of power. Its successor, the Prony brake, Fig. 7.1(b) also relied on mechanical friction and like the rope brake required cooling by water introduced into the hollow brake drum and removed by a scoop.

Both these devices are now obsolete. Their successors may be classified according to the means adopted for absorbing the mechanical power of the prime mover driving the dynamometer.

Classification of dynamometers

1. *Hydrokinetic or 'hydraulic' dynamometers* (*water brakes*). With the exception of the disc dynamometer, all machines work on similar principles, Fig. 7.2. A shaft carries a cylindrical rotor which revolves in a watertight casing. Toroidal recesses formed half in the rotor and half in the casing or stator are divided into pockets by radial vanes set at an angle to the axis of the rotor. When the rotor is driven centrifugal force sets up an intensive toroidal circulation as indicated by the arrows in Fig. 7.2(a). The effect is to transfer momentum from rotor to stator and hence to develop a torque resistant to the rotation of the shaft, balanced by an equal and opposite torque reaction on the casing.

A forced vortex of toroidal form is generated as a consequence of this motion, leading to high rates of turbulent shear in the water and the dissipation of power in the form of heat to the water. The centre of the vortex is vented to atmosphere by way of passages in the rotor and the virtue of the design is that power is absorbed with minimal damage to the moving surfaces, either from erosion or from the effects of cavitation.

The machines are of two kinds, depending on the means by which the resisting torque is varied.

1(a) *Constant fill machines: the classical Froude or sluice plate design*, Fig. 7.3 In this machine torque is varied by inserting or withdrawing pairs

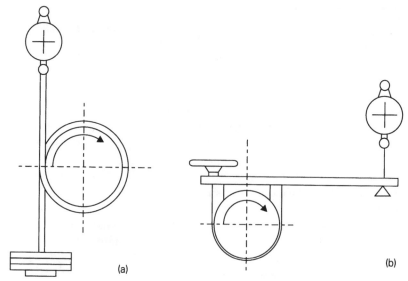

Figure 7.1.   *Early types of dynamometer: (a) rope brake; (b) Prony brake*

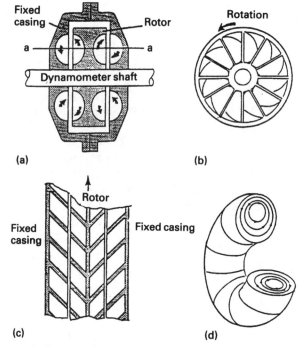

Figure 7.2.   *Hydrokinetic dynamometer: principle of operation: (a) section through dynamometer; (b) end view of rotor; (c) development of section a–a of rotor and casing; (d) representation of toroidal vortex*

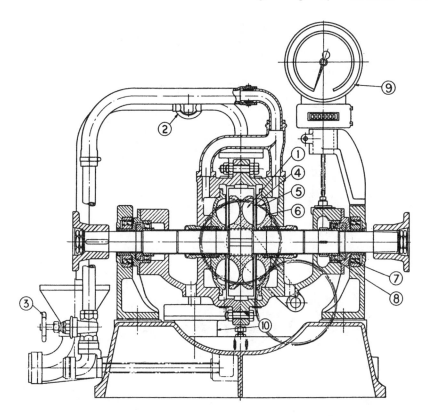

Typical cross-section through casing of Froude dynamometer, type DPX

(1) Rotor	(6) Casing liners
(2) Water outlet valve	(7) Casing trunnion bearing
(3) Water inlet valve	(8) Shaft bearing
(4) Sluice plates for load control	(9) Tachometer
(5) Water inlet holes in vanes	

Figure 7.3. *Froude sluice plate dynamometer*

of thin sluice plates between rotor and stator, thus controlling the extent of the development of the toroidal vortices.

1(b) *Variable fill machines*, Fig. 7.4 These machines are essentially similar to the Froude design but the sluice plates are omitted and the torque absorbed is varied by adjusting the quantity of water in the casing. This is achieved by a valve on the water outlet, associated with control systems of widely varying complexity. The particular advantage of the variable fill machine is that the

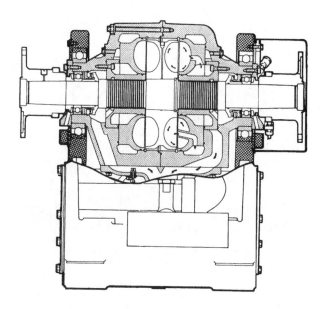

Figure 7.4.   *Variable fill hydraulic dynamometer*

torque may be varied much more rapidly than is the case with sluice plate control.

1(c) *'Bolt-on' variable fill machines*, Fig 7.5   These machines, available for many years in the USA, operate on the same principle as those described above, but are arranged to bolt directly on to the engine clutch housing or into the truck chassis. They do not require an elaborate test bed and are particularly useful when, for example, reconditioned engines are to be tested in moderate numbers. Machines are available for ratings up to about 1000 kW. In these machines load is controlled by an inlet control valve associated with a throttled outlet.

1(d) *Disc dynamometers*   These machines, not very widely used, consist of one or more flat discs located between flat stator plates, with a fairly small clearance. Power is absorbed by intensive shearing of the water and torque is controlled as in variable fill machines. Disc dynamometers have comparatively poor low speed performance but may be built to run at very high speeds, making them suitable for loading gas turbines. A variation is the perforated disc machine, in which there are holes in the rotor and stators, giving greater power dissipation for a given size of machine.

2. *Hydrostatic dynamometers*   Not very widely used, these machines consist generally of a combination of a fixed stroke and a variable stroke positive displacement multi-piston hydraulic pump/motor. The fixed stroke machine forms the dynamometer. An advantage of this arrangement is that, unlike most

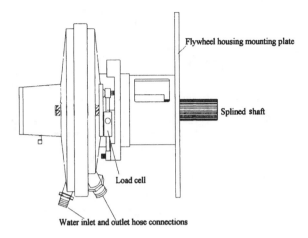

Flywheel housing mounting plate

Splined shaft

Load cell

Water inlet and outlet hose connections

Figure 7.5. *'Bolt-on' dynamometer*

other machines, it is capable of developing full torque down to zero speed and is also capable of acting as a source of power to 'motor' the engine under test.

3. *Electrical dynamometers*   The common feature of all these machines is that the power absorbed is transformed into electrical energy, either as power that is 'exported' from the machine or as eddy currents that in turn give rise to energy loss in the form of heat that is transferred to cooling water.

3(a) *Direct current (d.c.) dynamometers*, Fig. 7.6   These machines consist essentially of a trunnion mounted d.c. motor generator. Control is almost universally by means of a thyristor based a.c./d.c. converter. d.c. dynamometers are robust, easily controlled, and capable of motoring and starting as well as of absorbing power. Disadvantages include limited maximum speed and high inertia, which can present problems of torsional vibration, see Chapter 8, and limited rates of speed change.

3(b) *Alternating current (a.c.) dynamometers*   Developments in a.c. drive technology have permitted the evolution of a.c. machines of similar rating and performance to d.c. dynamometers, with the advantages of lower inertia and the absence of a commutator. These asynchronous machines consist essentially of an induction motor, the speed of which is controlled by varying the supply frequency. The power supply comprises a rectifier, an intermediate d.c. circuit and an inverter to produce the variable frequency supply. The machine is capable of regenerative braking.

3(c) *Eddy current dynamometers*, Fig. 7.7   These machines make use of the principle of electro-magnetic induction to develop torque and dissipate power. A toothed rotor of high-permeability steel rotates with a fine clearance between water-cooled steel loss plates. A magnetic field parallel to the machine axis is

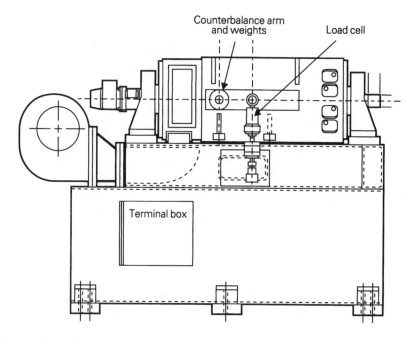

Figure 7.6.   *D.C. electrical dynamometer*

generated by two annular coils and motion of the rotor gives rise to changes in the distribution of magnetic flux in the loss plates.

This in turn gives rise to circulating eddy currents and the dissipation of power in the form of electrical resistive losses. Energy is transferred in the form of heat to cooling water circulating through passages in the loss plates, while some cooling is achieved by the radial flow of air in the gaps between rotor and plates.

Power is controlled by varying the current supplied to the annular exciting coils, and very rapid load changes are possible. Eddy current machines are simple and robust, the control system is simple and they are capable of developing substantial braking torque at quite low speeds. Unlike a.c. or d.c. dynamometers, however, they are unable to develop motoring torque.

Although no longer widely used, an alternative form of eddy current machine is also available. This employs a simple disc or drum design of rotor in which eddy currents are induced and the heat developed is transferred to water circulated through the gaps between rotor and stator. These 'wet gap' machines are liable to corrosion, have higher inertia, and have a high level of minimum torque, arising from drag of the cooling water in the gap.

4. *Friction dynamometers*, Fig. 7.8. These machines, in direct line of succession from the original rope brake, consist essentially of water-cooled multi-disc friction brakes. They are useful for low-speed applications, for

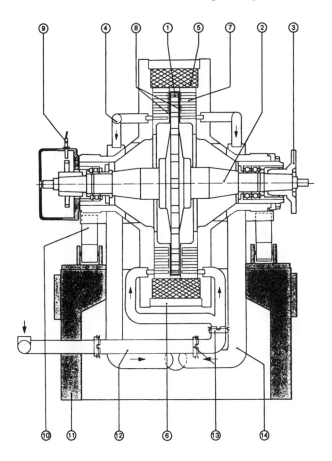

Figure 7.7. *Eddy current dynamometer. 1 Rotor, 2 rotor shaft, 3 coupling flange, 4 water outlet with thermostat, 5 excitation coil, 6 dynamometer housing, 7 cooling chamber, 8 air gap, 9 speed pick-up, 10 flexure support, 11 base, 12 water inlet, 13 joint, 14 water outlet pipe.*

example for measuring the power output of a vehicle transmission at the wheels, and have the advantage, shared with the hydrostatic dynamometer, of developing full torque down to zero speed.

5. *Air brake dynamometers* These devices, of which the Walker fan brake was the best-known example, are now largely obsolete. They consisted of a simple arrangement of radially adjustable paddles that imposed a torque that could be approximately estimated. They survive mainly for use in the field testing of helicopter engines, where high accuracy is not required and the noise is no disadvantage.

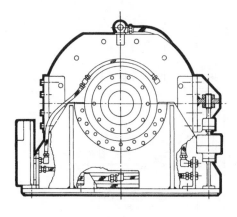
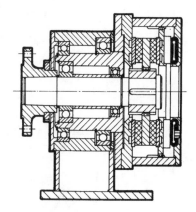

Figure 7.8.   *Friction brake*

Hybrid dynamometers

For completeness mention should be made of a combined design that is occasionally adopted for cost reasons. The d.c. or a.c. electrical dynamometer is capable of generating a motoring torque almost equal to its braking torque. However, the motoring torque required in engine testing seldom exceeds 30% of the engine power output. Since, for equal power absorption, a.c. and d.c. machines are more expensive than other types, it is sometimes worth while to run an electrical dynamometer in tandem with, for example, a variable fill hydraulic machine. Control of these machines is a more complex matter and the need to provide duplicate services, both electrical power and cooling water, is a further disadvantage. The solution may, however, on occasion be cost effective.

One, two or four quadrant?

Figure 7.9 illustrates diagrammatically the four 'quadrants' in which a dynamometer may be required to operate. Most engine testing takes place in the first quadrant, the engine running anticlockwise when viewed on the flywheel end. On comparatively rare occasions it is necessary for a test installation to accept engines running in either direction; one solution is to employ a dynamometer with couplings at both ends mounted on a turntable. Large marine engines are of course usually reversible.

All types of dynamometer are naturally able to run in the first (or second) quadrant. Hydraulic dynamometers are usually designed for one direction of rotation, though they may be run in reverse without damage. When designed specifically for bidirectional rotation they may be larger than a single-direction

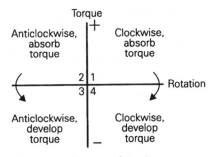

Figure 7.9. *Dynamometer operating quadrants*

machine of equivalent power. The torque measuring system must of course operate in both directions. Eddy current machines are inherently reversible.

When it is required to operate in the third and fourth quadrants (i.e. for the dynamometer to produce power as well as to absorb it) the choice is effectively limited to d.c. or a.c. machines, or to the hydrostatic or hybrid machine. These machines are generally reversible and therefore operate in all four quadrants.

There is an increasing requirement for four-quadrant operation as a result of the growth in transient testing, see Chapter 15, with its call for very rapid load changes and even for torque reversals. This presents certain problems, as d.c. and a.c. machines have higher rotational inertia than other types, and this militates against rapid speed changes.

If mechanical losses in the engine are to be measured by 'motoring', see Chapter 12, a four-quadrant machine is required, and a useful feature of such a machine is its ability also to start the engine. Table 7.1 summarizes the performance of machines in this respect.

Table 7.1. *Operating quadrants (see Fig. 7.9)*

Type of machine	Quadrant
*Hydraulic sluice plate	1 or 2
*Variable fill hydraulic	1 or 2
*'Bolt On' variable fill hydraulic	1 or 2
Disc type hydraulic	1 and 2
Hydrostatic	1, 2, 3, 4
d.c. electrical	1, 2, 3, 4
a.c. electrical	1, 2, 3, 4
Eddy current	1 and 2
Friction brake	1 and 2
Air brake	1 and 2
Hybrid	1, 2, 3, 4

* These machines are also available for one- and two-quadrant operation.

Matching engine and dynamometer characteristics

The different types of dynamometer have significantly different torque-speed and power–speed curves, and this can affect the choice made for a given application. Figure 7.10 shows the performance curves of a typical hydraulic dynamometer. The different elements of the performance envelope are as follows:

(a) Dynamometer full (or sluice plates wide open). Torque increases with square of speed, no torque at rest.
(b) Performance limited by maximum permitted shaft torque.
(c) Performance limited by maximum permitted power, which is a function of cooling water through-put and its maximum permitted temperature rise.
(d) Maximum permitted speed.
(e) Minimum torque corresponding to minimum permitted water flow.

Figure 7.11 shows the considerably different performance envelope of an electrical machine, made up of the following elements:

(a) Constant torque corresponding to maximum current and excitation.
(b) Performance limited by maximum permitted power output of machine.
(c) Maximum permitted speed.

Since these are 'four-quadrant' machines, power absorbed can be reduced to zero and there is no minimum torque curve.
Figure 7.12 shows the performance curves for an eddy current machine, which lie between those of the previous two machines:

(a) Low speed torque corresponding to maximum permitted excitation.
(b) Performance limited by maximum permitted shaft torque.

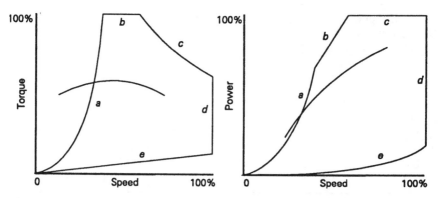

Figure 7.10. *Performance curves: hydraulic dynamometer*

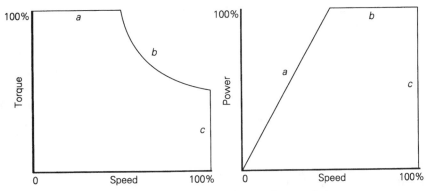

Figure 7.11. *Performance curves: d.c. or a.c. electrical dynamometer*

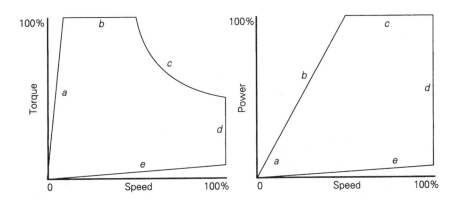

Figure 7.12. *Performance curves eddy current dynamometer*

(c) Performance limited by maximum permitted power, which is a function of cooling water through-put and its maximum permitted temperature rise.
(d) Maximum permitted speed.
(e) Minimum torque corresponding to residual magnetization, windage and friction.

In choosing a dynamometer for an engine or range of engines it is essential to superimpose the maximum torque– and power–speed curves onto the dynamometer envelope. See the example in Fig. 7.10, which demonstrates a typical problem: the hydraulic machine is incapable of developing sufficient torque at the bottom end of the speed range.

For best accuracy it is desirable to choose the smallest machine that will cope with the largest engine to be tested. Hydraulic dynamometers are generally able to deal with a moderate degree of overload and overspeed, but it is undesirable to run electrical machines beyond their rated limits: this can lead to damage to commutators, overheating and distortion of eddy current loss plates.

Careful attention must also be given to the arrangements for coupling engine and dynamometer, see Chapter 8.

Engine starting and cranking

Starting an engine when it is connected to a dynamometer may present the cell designer and operator with a number of problems, and is a factor to be borne in mind when selecting the dynamometer. If the engine is fitted with a starter motor the cell system must provide the d.c. supply and associated switching; in the absence of a starter a complete system to start and crank the engine must be available.

Engine cranking, no starter motor

The cell cranking system must be capable of accelerating the engine to its normal starting speed and, in some cases, of disengaging when the engine fires. A four-quadrant dynamometer will be capable of starting the engine directly. The power available from any four- quadrant machine will always be greater than that required and excessive starting torque must be avoided by an alarm system that monitors starting current, otherwise an engine locked by seizure or fluid in a cylinder, may cause damage to the drive line.

The preferred method of providing other types of dynamometer with a starting system is to mount an electric motor at the non-engine end of the dynamometer shaft, driving through an over-running or remotely engaged clutch, and generally through a speed-reducing belt drive. The clutch half containing the mechanism should be on the input side, otherwise it will be affected by the torsional vibrations usually experienced by dynamometer shafts. The motor may be mounted above or below the dynamometer to save length.

The sizing of the motor must take into account the maximum break-away torque expected, usually estimated as twice the average cranking torque, while the normal running speed of the motor should correspond to the desired cranking speed. The choice of motor and associated starter must take into account the maximum number of starts per hour that may be required, both in normal use and when dealing with a faulty engine. The running regime of the motor is demanding, involving repeated bursts at overload, with the intervening time at rest, and an independent cooling fan may be necessary.

Some modern diesel engines, when 'green'*, require cranking at more than the normal starting speed, sometimes as high as 1200 rev/min, in order to prime the fuel system. In such cases a two-speed or fully variable speed starter motor may be necessary.

The system must be designed to impose the minimum parasitic torque when disengaged, since this torque will not be sensed by the dynamometer measuring system.

In some cases, to avoid this source of inaccuracy, the motor may be mounted directly on the dynamometer carcase and permanently coupled to the dynamometer shaft by a belt drive. This imposes an additional load on the trunnion bearings, which may lead to brinelling, and it also increases the effective moment of inertia of the dynamometer. However it has the advantage that motoring and starting torque may be measured by the dynamometer system.

An alternative solution is to use a standard vehicle engine starter motor in conjunction with a gear ring carried by a 'dummy flywheel' carried on a shaft with separate bearings incorporated in the drive line, but this may have the disadvantage of complicating the torsional behaviour of the system.

Engine mounted starter systems

If the engine is fitted with its own starter motor on arrival at the test stand all that must be provided is the necessary 12 V or 24 V supply; the traditional approach has been to locate a suitable battery as close as possible to the starter motor, with a suitable battery charger supply. This system is not ideal, as the battery needs to be in a suitably ventilated box, to avoid the risk of accidental shorting, and will take up valuable space. Special transformer/rectifier units designed to replace batteries for this duty are on the market. They will include an 'electrical services box' to provide power in addition for ignition systems and diesel glow plugs. In large integrated systems there will be a busbar system for the d.c. supplies.

The engine starter will be presented with a situation not encountered in normal service: it will be required to accelerate the whole dynamometer system in addition to the engine while a 'green' engine may exhibit a very high breakaway torque and require prolonged cranking at high speed before it fires.

---

* A green engine is one that has never been run. The rubbing surfaces may be dry, the fuel system may need priming, and there is always the possibility that it may be defective and incapable of running.

Non-electrical starting systems

Diesel engines larger than the automotive range are usually started by means of compressed air, admitted to the cylinders by way of starting valves. In some cases it is necessary to move the crankshaft to the correct starting position, either by barring or using an engine mounted inching motor. The test facility should include a compressor and a receiver of capacity at least as large as that recommended for the engine in service.

Compressed air or hydraulic motors are sometimes used instead of electric motors to provide cranking power but have no obvious advantages over a d.c. electric motor, apart from a marginally reduced fire risk in the case of compressed air, provided the supply is shut off automatically in the case of fire.

In Chapter 8, p.156, attention is drawn to the possibility of overloading flexible couplings in the drive line during the starting process, and particularly when the engine first fires. This should not be overlooked.

The use of calibrated generators

A means of absorbing power sometimes worth consideration is the use of a calibrated a.c. or d.c. generator. The computing power necessary to calculate

Table 7.2. *Availability of dynamometers for different applications*

	Small engines up to 50 kW	*Automative engines 50–500 kW*	*Medium speed marine and stationary 500–5000 kW*	*Large marine 5000– 50 000 kW*	*Gas turbine*
Hydraulic	A	A	A	A	A
Bolt on hydraulic	A	A	NA	NA	NA
Disc hydraulic	NA	NA	NA	NA	A
Hydrostatic	NA	B	NA	NA	NA
d.c. electrical	A	A	A	NA	NA
a.c. electrical	B	A	NA	NA	NA
Eddy current	A	A	NA	NA	B
Eddy current (wet gap)	NA	B	NA	NA	NA
Airbrake	B	NA	NA	NA	B
Hybrid	NA	B	NA	NA	NA

A – preferred choice
B – alternative
NA – not applicable

Table 7.3. *Dynamometers: advantages and disadvantages*

Dynamometer type	Advantages	Disadvantages
Froude sluice plate	Obsolete but many cheap reconditioned units available Robust Tolerant of overload	Slow response to change in load demand Manual load control not easily automated
Variable fill machines	Capable of rapid load change, control readily automated Robust Tolerant of overload Available for all power ranges up to 10 000 kW	'Open' water system required Can suffer from cavitation corrosion or corrosion damage
'Bolt on' variable fill hydraulic	Cheap and do not require elaborate installation Up to 1000 kW	Lower accuracy of measurment and control than fixed base machine
Disc type hydraulic	Suitable for high speeds	Poor low speed performance
Hydrostatic	For special purposes, give complete control of speed in all four quadrants	Noisy, expensive. Use of high pressure oil
d.c. electrical	Capable of rapid load change. Well adapted to computer control. Four quadrant. No cooling water required	Expensive. High rotational inertia
a.c. electrical	As d.c. with advantage of lower inertia	Expensive. Possible problems with 'cogging'
Eddy current	Capable of rapid load change. Well adapted to computer control. Simple and robust. Low inertia	No motoring facility. Vulnerable to poor cooling supply. Not capable of sustaining overload
Friction brake	Useful for high-torque low-speed applications	Tendency to wear
Air brake	Cheap	Noisy, limited accuracy
Hybrid	Possible cost advantage over pure d.c. or a.c. machines	Complex control system

the torque from measurements of speed, volts and amps at terminals, and power factor is now quite cheaply available, but doubts will remain regarding accuracy. In the view of the authors such machines, which certainly form an economical method of absorbing power, will probably be fitted with flange mounted torque transducers, Fig.9.5. Appropriate fields of application are as listed for d.c. electrical and a.c. electrical machines in Tables 7.2 and 7.3.

Choice of dynamometer

Table 7.2 lists the various types of dynamometer and indicates their applicability for various classes of engine. In most cases several choices are available and it will be necessary to consider the special features of each type of dynamometer and to evaluate the relative importance of these in the particular case. These features are listed in Table 7.3, and other special factors are considered later.

Some additional considerations

The final choice of dynamometer for a given application may be influenced by some of the following factors:

(1) Load factor. If the machine will spend long periods out of use the possibilities of corrosion must be considered, particularly in the case of hydraulic or wet gap eddy current machines. Can the machine be drained readily? Should the use of corrosion inhibitors be considered?

(2) Overloads. If it may be necessary to consider occasional overloading of the machine a hydraulic machine may be preferable, in view of its greater tolerance of such conditions. Check that the torque measuring system has adequate capacity.

(3) Large and frequent changes in load. This can give rise to problems with eddy current machines, due to expansion and contraction with possible distortion of the loss plates.

(4) Wide range of engine sizes to be tested. It may be difficult to achieve good control and adequate accuracy when testing the smallest engines, while the minimum dynamometer torque may also be inconveniently high.

(5) How are engines to be started? If a non-motoring dynamometer is favoured it may be necessary to fit a separate starter to the dynamometer shaft. This represents an additional maintenance commitment and may increase inertia.

(6) Is there an adequate supply of cooling water of satisfactory quality? Hard water will result in blocked cooling passages and some water treatments can give rise to corrosion. This may be a good reason for choosing d.c. or a.c. dynamometers, despite extra cost.

(7) Is the pressure of the water supply subject to sudden variations? Sudden pressure changes or regular pulsations will affect the stability of control of hydraulic dynamometers. Eddy current and indirectly cooled machines are unaffected providing pressure does not fall below emergency trip levels.

(8) Is the electrical supply voltage liable to vary as the result of other loads on the same circuit? With the exception of air brakes and manually controlled hydraulic machines all dynamometers are affected by electrical interference and voltage changes.

(9) Is it proposed to use a shaft docking system for coupling engine and dynamometer? Are there any special features or heavy overhung or axial loads associated with the coupling system? Such features should be discussed with the dynamometer manufacturer before making a decision. Some machines, notably the Schenck flexure plate mounting system, are not suited to taking axial loads.

Finally, and perhaps most important, account should be taken of the level of experience and technical ability of the testing staff. Technicians used to simple hydraulic brakes may have difficulty in mastering the operation of an advanced computer-controlled electrical machine.

The supplier of any new dynamometer should offer an acceptance test and basic training in operation, calibration and safety of the new machine. A careful check should be made on the level of technical support, including availability of spares and local service facilities, offered by the manufacturer.

Summary

It is evident that the choice of the best dynamometer for a given application is by no means a simple matter. A logical progression might be:

(1) Decide whether two- or four-quadrant capability is required, Table 7.1.
(2) List the available choices from Table 7.2.
(3) Consider the various advantages and disadvantages listed in Table 7.3.
(4) Work through the sequence 'Matching engine and dynamometer characteristics'.
(5) Check through the 'additional considerations'.
(6) Finally, think carefully about the questions raised in the last paragraph above.

It may be necessary to go through this procedure two or three times.

Further reading

BS 4575 Parts 1 to 3, *Fluid Power Transmission and Control Systems.*

8 Coupling the engine to the dynamometer

The selection of suitable couplings and shaft for the connection of the engine to the dynamometer is by no means a simple matter. Incorrect choice or faulty design may give rise to a number of problems:

- torsional oscillations
- vibration of engine or dynamometer
- whirling of coupling shaft
- damage to engine or dynamometer bearings
- excessive wear of shaft line components
- catastrophic failure of coupling shafts
- starting problems.

This whole subject, the coupling of engine and dynamometer, can give rise to more trouble than any other routine aspect of engine testing, and a clear understanding of the many factors involved is desirable. The following chapter covers all the main factors, but in some areas more extensive analysis may be necessary and appropriate references are given.

The nature of the problem

The special feature of the problem is that it must be considered afresh each time an engine not previously encountered is installed. It must also be recognized that unsatisfactory torsional behaviour is associated with the whole system – engine, coupling shaft and dynamometer – rather than with the individual components, all of which may be quite satisfactory in themselves.

Problems arise partly because the dynamometer is seldom equivalent dynamically to the system driven by the engine in service. This is particularly the case with vehicle engines. In the case of a vehicle with rear axle drive, the driveline consists of a clutch, which may itself act as a torsional damper, followed by a gearbox, the characteristics of which are low inertia and some damping capacity. This is followed by a drive shaft and differential, itself having appreciable damping, two half – shafts and two wheels, both with

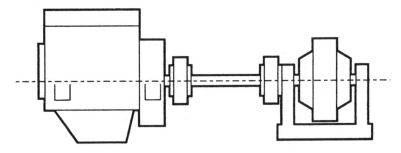

Figure 8.1.   *Engine and dynamometer drive line*

substantial damping capacity and running at much slower speed than the engine, thus reducing their effective inertia.

When coupled to a dynamometer, Fig. 8.1, this system, with its built-in damping and moderate inertia, is replaced by a single drive shaft connected to a single rotating mass, the dynamometer, running at the same speed as the engine. The clutch may or may not be retained.

Particular care is necessary where the moment of inertia of the dynamometer is more than about twice that of the engine. A further consideration that must be taken seriously concerns the effect of the difference between the engine mounting arrangements in the vehicle and on the testbed. This can lead to vibrations of the whole engine that can have a disastrous effect on the drive shaft, see Chapter 5.

Background reading

The mathematics of the subject are complex and not readily accessible. Den Hartog[1] gives what is in the authors' view the clearest exposition of fundamentals. Ker Wilson's classical treatment in five volumes[2] is probably still the best source of comprehensive information; his abbreviated version[3] is sufficient for most purposes. M.E.P. have published a useful practical handbook[4] while Lloyd's Register[5] give rules for the design of marine drives that are also useful in the present context.

Torsional oscillations and critical speeds

In its simplest form, the engine-dynamometer system may be regarded as equivalent to two rotating masses connected by a flexible shaft, Fig. 8.2. Such a system has an inherent tendency to develop torsional oscillations. The two masses can vibrate 180° out of phase about a node located at some point along the shaft between them. The oscillatory movement is superimposed on any

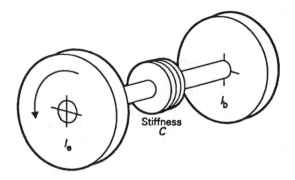

Figure 8.2.   *Two-mass system*

steady rotation of the shaft. The resonant or critical frequency of torsional oscillation of this system is given by:

$$n_c = \frac{60}{2\pi} \sqrt{\frac{C_c(I_e + I_b)}{I_e I_b}} \tag{1}$$

If an undamped system of this kind is subjected to an exciting torque of constant amplitude T_{ex} and frequency n, the relation between the amplitude of the induced oscillation θ and the ratio n/n_c is as shown in Fig 8.3. At low frequencies the combined amplitude of the two masses is equal to the static deflection of the shaft under the influence of the exciting torque, $\theta_0 = T_{ex}/C_s$.

As the frequency increases the amplitude rises and at $n = n_c$ it becomes theoretically infinite: the shaft may fracture or non-linearities and internal damping may prevent actual failure. With further increases in frequency the amplitude falls and at $n = \sqrt{2}n_c$ it is down to the level of the static deflection. Amplitude continues to fall with increasing frequency.

The shaft connecting engine and dynamometer must be designed with a suitable stiffness C_s to ensure that the critical frequency lies outside the normal operating range of the engine, and also with a suitable degree of damping to ensure that the unit may be run through the critical speed without the development of a dangerous level of torsional oscillation. Fig. 8.3 also shows the behaviour of a damped system. The ratio θ/θ_0 is known as the dynamic magnifier M. Of particular importance is the value of the dynamic magnifier at the critical frequency, M_c. The curve of Fig. 8.3 corresponds to a value $M_c = 5$.

Torsional oscillations are excited by the variations in engine torque associated with the pressure cycles in the individual cylinders (also, though usually of less importance, by the variations associated with the movement of the reciprocating components).

Fig. 8.4 shows the variation in the case of a typical single cylinder four-

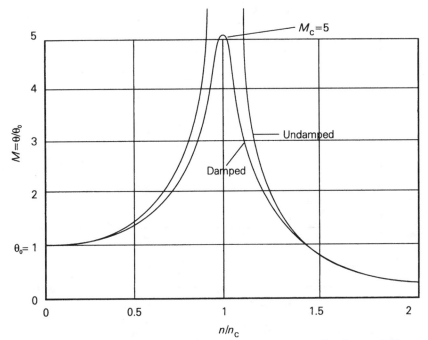

Figure 8.3. *Relation between frequency ratio, amplitude and dynamic magnifier M*

stroke diesel engine. It is well known that any periodic curve of this kind may be synthesized from a series of *harmonic components*, each a pure sine wave of a different amplitude having a frequency corresponding to a multiple or sub-multiple of the engine speed and the figure shows the first six components.

The *order* of the harmonic defines this multiple. Thus a component of order $N_o = \frac{1}{2}$ occupies two revolutions of the engine, $N_o = 1$ one revolution and so on. In the case of a four cylinder four-stroke engine there are two firing strokes per revolution of the crankshaft and the turning moment curve of Fig. 8.4 is repeated at intervals of 180°. In a multi-cylinder engine the harmonic components of a given order for the individual cylinders are combined by representing each component by a vector, in the manner illustrated on p.72, Fig. 5.3, for the inertia forces. A complete treatment of this process is beyond the scope of this book, but the most significant results may be summarized as follows.

The first major critical speed for a multi-cylinder in-line engine is of order:

$$N_o = N_{CYL}/2 \text{ for a four-stroke engine} \tag{2a}$$

$$N_o = N_{CYL} \text{ for a two-stroke engine} \tag{2b}$$

Thus, in the case of a four cylinder four-stroke engine the major critical speeds

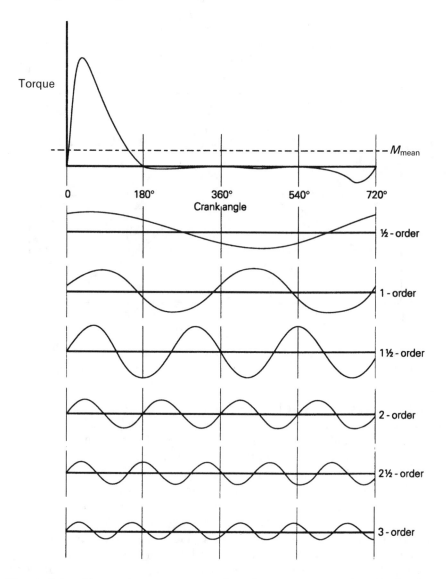

Figure 8.4. *Harmonic components of turning moment, single cylinder four-stroke gasolene engine*

are of order 2, 4, 6, etc. In the case of a six cylinder engine they are of order 3, 6, 9, etc.

The distinction between a major and a minor critical speed is that in the case of an engine having an infinitely rigid crankshaft it is only at the major critical speeds that torsional oscillations can be induced. This, however, by no means

Table 8.1. *p factors*

Order	$\frac{1}{2}$	1	$1\frac{1}{2}$	2	$2\frac{1}{2}$	3...	8
p factor	2.16	2.32	2.23	1.91	1.57	1.28	0.08

implies that in large engines having a large number of cylinders the minor critical speeds may be ignored.

At the major critical speeds the exciting torques T_{ex} of all the individual cylinders in one line act in phase and are thus additive (special rules apply governing the calculation of the combined excitation torques for Vee engines).

The first harmonic is generally of most significance in the excitation of torsional oscillations, and for engines of moderate size, such as passenger vehicle engines, it is generally sufficient to calculate the critical frequency from eq. (1), then to calculate the corresponding engine speed from:

$$N_c = n_c / N_o \tag{3}$$

The stiffness of the connecting shaft between engine and dynamometer should be chosen to that this speed does not lie within the range in which the engine is required to develop power.

In the case of large multi-cylinder engines the 'wind up' of the crankshaft as a result of torsional oscillations can be very significant and the two-mass approximation is inadequate; in particular the critical speed may be over-estimated by as much as 20% and more elaborate analysis is necessary. The subject is dealt with in several different ways in the literature; perhaps the easiest to follow is that of Den Hartog[1]. The starting point is the value of the mean turning moment developed by the cylinder, M_{mean}, Fig. 8.4. Values are given for a so-called 'p factor', by which M_{mean} is multiplied to give the amplitude of the various harmonic excitation forces. Table 8.1, reproduced from ref. 1, shows typical figures for a 4-stroke medium speed diesel engine. Exciting torque:

$$T_{ex} = p. \, M_{mean} \tag{4}$$

The relation between M_{mean} and imep (indicated **mean effective pressure**) is given by:

$$\text{for a 4-stroke engine} \quad M_{mean} = p_i \cdot \frac{B^2 S}{16} \cdot 10^{-4} \tag{5a}$$

$$\text{for 2-stroke engine} \quad M_{mean} = p_i \cdot \frac{B^2 S}{8} \cdot 10^{-4} \tag{5b}$$

Lloyd's Rulebook[5], the main source of data on this subject, expresses the amplitude of the harmonic components rather differently, in terms of a 'component of tangential effort', T_m. This is a pressure that is assumed to act upon the piston at the crank radius $S/2$. Then exciting torque per cylinder:

$$T_{ex} = T_m \frac{\pi B^2}{4} \frac{S}{2} \times 10^{-4} \tag{6}$$

Lloyd's give curves of T_m in terms of the indicated mean effective pressure p_i, and it may be shown that the values so obtained agree closely with those derived from Table 8.1.

The amplitude of the vibratory torque T_v induced in the connecting shaft by the vector sum of the exciting torques for all the cylinders, $\sum T_{ex}$, is given by:

$$T_v = \frac{\sum T_{ex} M_c}{(1 + I_e/I_b)} \tag{7}$$

The complete analysis of the torsional behaviour of a multi-cylinder engine is a substantial task, though computer programmes are available which reduce the effort required. As a typical example, Figure 8.5 shows the 'normal elastic curves' for the first and second modes of torsional oscillation of a 16-cylinder Vee engine coupled to a hydraulic dynamometer. These curves show the amplitude of the torsional oscillations of the various components, relative to that at the dynamometer which is taken as unity. The natural frequencies are respectively $n_c = 4820$ c.p.m. and $n_c = 6320$ c.p.m. The curves form the basis for further calculations of the energy input giving rise to the oscillation. In the case of the engine under consideration these showed a very severe fourth order oscillation, $N_o = 4$, in the first mode. (For an engine having eight cylinders in line the first major critical speed, from eq. (2a), is of order $N_o = 4$). The engine speed corresponding to the critical frequency of torsional oscillation is given by:

$$N_c = n_c/N_o \tag{8}$$

giving, in the present case, $N_c = 1205$ rev/min, well within the operating speed range of the engine. Further calculations showed a large input of oscillatory energy at $N_o = 4\frac{1}{2}$, a minor critical speed, in the second mode, corresponding to a critical engine speed of $6320/4\frac{1}{4} = 1404$ rev/min, again within the operating range. Several failures of the shaft connecting engine and dynamometer occurred before a safe solution was arrived at.

This example illustrates the need for caution and for full investigation in setting up large engines on the test bed.

It is not always possible to avoid running close to or at critical speeds, and this situation is usually dealt with by the provision of *torsional vibration dampers*, in which the energy fed into the system by the exciting forces is absorbed by viscous shearing. Such dampers are commonly fitted at the non-flywheel end of engine crankshafts. In some cases it may also be necessary to consider their use as a component of engine test cell drive lines, when they are located either as close as possible to the engine flywheel, or at the dynamometer. The damper must be 'tuned' to be most effective at the critical frequency and the selection of a suitable damper involves equating the energy

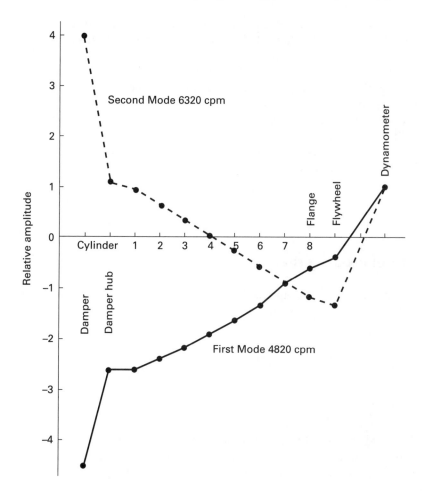

Figure 8.5. *Normal elastic curves for 16-cylinder V engine coupled to hydraulic dynamometer*

fed into the system per cycle with the energy absorbed by viscous shear in the damper. This leads to an estimate of the magnitude of the oscillatory stresses at the critical speed. For a clear treatment of the theory see Den Hartog[1].

Points to remember:

- as a general rule, it is good practice to avoid running the engine under power at speeds between 0.8 and 1.2 × critical speed. If it is necessary to take the engine through the critical speed this should be done off load and as quickly as possible. With high inertia dynamometers the transient vibratory torque may well exceed the mechanical capacity of the drive line and the margin of safety of the drive line components may need to be increased

- problems frequently arise when the inertia of the dynamometer much exceeds that of the engine: a detailed torsional analysis is desirable when this factor exceeds 2. This situation usually occurs when it is found necessary to run an engine of much smaller output than the rated capacity of the dynamometer

- the simple 'two mass' approximation of the engine–dynamometer system is inadequate for large engines and may lead to over-estimation of the critical speed.

Design of coupling shafts

The maximum shear stress induced in a shaft, diameter D, by a torque T Nm is given by:

$$\tau = \frac{16T}{\pi D^3} \quad Pa \tag{9a}$$

in the case of a tubular shaft, bore diameter d, this becomes:

$$\tau = \frac{16TD}{\pi(D^4 - d^4)} \tag{9b}$$

For steels, the shear yield stress is usually taken as equal to $0.75 \times$ yield stress in tension. A typical choice of material would be a nickel–chromium–molybdenum steel, to specification BS 817M40, (previously En 24), heat treated to the T condition.

The various stress levels for this steel are roughly as follows:

ultimate tensile strength	not less than 850 MPa, (55 t.s.i.)
0.1% proof stress in tension	550 MPa
ultimate shear strength	500 MPa
0.1% proof stress in shear	300 MPa
shear fatigue limit in reversed stress	± 200 MPa

It is clear that the permissible level of stress in the shaft will be a small fraction of the ultimate tensile strength of the material.

The choice of designed stress level at the maximum rated steady torque is influenced by two principal factors:

- stress concentrations. For a full treatment of this very important subject see ref. (6). There are two particularly critical locations to be considered:

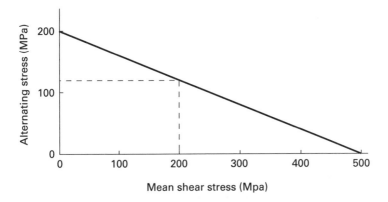

Figure 8.6.   *Goodman diagram, steel shaft in shear*

at a shoulder fillet, such as is provided at the junction with an integral flange. For a ratio fillet radius/shaft diameter = 0.1 the stress concentration factor is about 1.4, falling to 1.2 for r/d = 0.2.
At the semicircular end of a typical rectangular keyway the stress concentration factor reaches a maximum of about 2.5 × nominal shear stress at an angle of about 50° from the shaft axis. The authors have seen a number of failures at this location.

• cyclic stresses associated with torsional oscillations. As, even in the most carefully designed installation involving an internal combustion engine, some torsional oscillation will be present, it is wise to select a conservative value for the nominal (steady state) shear stress in the shaft.

Stress concentration factors apply to the cyclic stresses rather than to the steady state stresses. Figure 8.6 shows diagrammatically the Goodman diagram[7] for a steel having the properties specified above. This diagram indicates approximately the relation between the steady shear stress and the permissible oscillatory stress. The example shown indicates that, for a steady torsional stress of 200 MPa, the accompanying oscillatory stress (calculated after taking into account any stress concentration factors) should not exceed ±120 MPa. In the absence of detailed design data, it is good practice to design shafts for use in engine test beds very conservatively, since the consequences of shaft failure can be so serious. A shear stress calculated in accordance with eq. (9) of about 100 MPa for a steel with the properties listed should be safe under all but the most unfavourable conditions. To put this in perspective, a shaft 100 mm diameter. designed on this basis would imply a torque of 19 600 Nm, or a power of 3100 kW at 1500 rev/min.

The torsional stiffness of a solid shaft of diameter D and length L is given by:

$$C_S = \frac{\pi D^4 G}{32L} \tag{10a}$$

for a tubular shaft, bore d:

$$C_S = \frac{\pi(D^4 - d^4)}{32L} \tag{10b}$$

Shaft whirl

The coupling shaft is usually supported at each end by a universal joint or flexible coupling. Such a shaft will 'whirl' at a rotational speed N_W (also at certain higher speeds in the ratio $2^2 N_W, 3^2 N_W$, etc.).

The whirling speed of a solid shaft of length L is given by:

$$N_W = \frac{30\pi}{L^2} \sqrt{\frac{E\pi D^4}{64 W_s}} \tag{11}$$

It is desirable to limit the maximum engine speed to about $0.8 N_W$. However when using rubber flexible couplings it is essential to allow for the radial flexibility of these couplings, since this can drastically reduce the whirling speed. *It is sometimes the practice to fit self-aligning rigid steady bearings at the centre of flexible couplings in high-speed applications, but these are liable to give fretting problems and are not universally favoured.*

As is well known, the whirling speed of a shaft is identical with its natural frequency of transverse oscillation. To allow for the effect of transverse coupling flexibility the simplest procedure is to calculate the transverse critical frequency of the shaft plus two half couplings from the equation:

$$N_t = \frac{30}{\pi} \sqrt{\frac{k}{W}} \tag{12a}$$

where $W =$ mass of shaft + half couplings and $k =$ combined radial stiffness of the two couplings.

Then whirling speed N taking this effect into account will be given by:

$$\left(\frac{1}{N}\right)^2 = \left(\frac{1}{N_W}\right)^2 + \left(\frac{1}{N_t}\right)^2 \tag{12b}$$

Couplings

The choice of the appropriate coupling for a given application is not easy: the majority of drive line problems probably have their origin in an incorrect choice of components for the drive line, and are usually cured by changes in this region. A complete discussion would much exceed the scope of this book, but the reader concerned with drive line design should obtain a copy of ref. 4, which gives a comprehensive treatment together with a valuable procedure for selecting the best type of coupling for a given application. A very brief summary of the main types of coupling follows.

Quill shaft with integral flanges and rigid couplings

This type of connection is best suited to the situation where a driven machine is permanently coupled to the source of power, when it can prove to be a simple and reliable solution. It is not well suited to test bed use, since it is intolerant of relative vibration and misalignment.

Quill shaft with toothed or gear type couplings

Gear couplings are very suitable for high powers and speeds, and can deal with relative vibration and some degree of misalignment, but this must be very carefully controlled to avoid problems of wear and lubrication. Such shafts are inherently stiff in torsion.

Conventional 'cardan shaft' with universal joints

These shafts are readily available from a number of suppliers, and are probably the preferred solution in the majority of cases. However standard automotive type shafts can give trouble when run at speeds in excess of those encountered in vehicle applications. A correct 'built-in' degree of misalignment is necessary to avoid fretting of the needle rollers.

Multiple membrane couplings

These couplings, Fig 8.7, are stiff in torsion but tolerant of a moderate degree of misalignment and relative axial displacement. They can be used for very high speeds.

Elastomeric element couplings

There is a vast number of different designs on the market and selection is not easy. Ref. 8 may be found helpful. The great advantage of these couplings is that their torsional stiffness may be varied widely by changing the elastic elements and problems associated with torsional vibrations and critical speeds dealt with, see the next section.

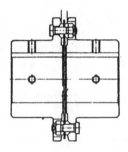

Figure 8.7.   *Multiple steel disc type flexible coupling*

Damping: the role of the flexible coupling

The earlier discussion leads to two main conclusions: the engine–dynamometer system is susceptible to torsional oscillations and the internal combustion engine is a powerful source of forces calculated to excite such oscillations. The magnitude of these undesirable disturbances in any given system is a function of the damping capacities of the various elements: the shaft, the couplings, the dynamometer and the engine itself.

The couplings are the only element of the system the damping capacity of which may readily be changed, and in many cases, for example with engines of automotive size, the damping capacity of the remainder of the system may be neglected, at least in an elementary treatment of the problem, such as will be given here.

The dynamic magnifier M, Fig. 8.3, has already been mentioned as a measure of the susceptibility of the engine–dynamometer system to torsional oscillation. Now referring to Fig. 8.1, let us assume that there are two identical flexible couplings, of stiffness C_c, one at each end of the shaft, and that these are the only sources of damping. Fig. 8.8 shows a typical torsionally resilient coupling in which torque is transmitted by way of a number of shaped rubber blocks or bushes which provide torsional flexibility, damping and a capacity to take up misalignment.

The torsional characteristics of such a coupling are shown in Fig. 8.9. These differ in three important respects from those of, say, a steel shaft in torsion:

- the coupling does not obey Hooke's Law: the stiffness or coupling rate $C_c = \Delta T / \Delta \theta$ increases with torque. This is partly an inherent property of the rubber and partly a consequence of the way it is constrained
- the shape of the torque–deflection curve is not independent of frequency. Dynamic torsional characteristics are usually given for a cyclic frequency of 10 Hz. If the load is applied slowly the stiffness is found to be substantially less. The following values of the ratio dynamic stiffness (at 10 Hz) to static stiffness of natural rubber of varying hardness are taken from ref. 4.

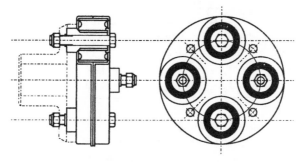

Figure 8.8. *Rubber bush type torsionally resilient coupling*

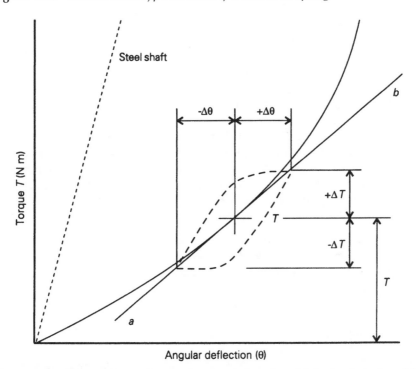

Figure 8.9. *Dynamic torsional characteristic of multiple bush type rubber coupling*

Shore (IHRD) Hardness	40	50	60	70
$\dfrac{\text{Dynamic stiffness}}{\text{Static stiffness}}$	1.5	1.8	2.1	2.4

Since the value of C_c varies with the deflection, manufacturers usually quote a single figure which corresponds to the slope of the tangent *ab* to the

Table 8.2.   *Damping energy ratio ψ*

Shore (IHRD) hardness	50/55	60/65	70/75	75/80
Natural rubber	0.45	0.52	0.70	0.90
Neoprene		0.79		
Styrene-butadiene (SBR)		0.90		

Table 8.3.   *Dynamic magnifier M*

Shore (IHRD) hardness	50/55	60/65	70/75	75/80
Natural rubber	10.5	8.6	5.2	2.7
Neoprene		4.0		
Styrene-butadiene (SBR)		2.7		

torque–deflection curve at the rated torque, typically one third of the max-imum permitted torque.

- if a cyclic torque $\pm\Delta T$, such as that corresponding to a torsional vibration, is superimposed on a steady torque T, Fig. 8.9, the deflection follows a path similar to that shown dotted. It is this feature, the hysteresis loop, which results in the dissipation of energy, by an amount ΔW proportional to the area of the loop, that is responsible for the damping characteristics of the coupling.

Damping energy dissipated in this way appears as heat in the rubber and can, under adverse circumstances, lead to overheating and rapid destruction of the elements. The appearance of rubber dust inside coupling guards is a warning sign.

The damping capacity of a component such as a rubber coupling is described by the *damping energy ratio*:

$$\psi = \frac{\Delta W}{W}$$

This may be regarded as the ratio of the energy dissipated by hysteresis in a single cycle to the elastic energy corresponding to the wind-up of the coupling at mean deflection:

$$W = \tfrac{1}{2}T\theta = \tfrac{1}{2}T^2/C_c$$

The damping energy ratio is a property of the rubber. Some typical values are given in Table 8.2. The dynamic magnifier is a function of the damping energy ratio: as would be expected a high damping energy ratio corresponds to a low dynamic magnifier. Some authorities give the relation:

$$M = 2\pi/\psi$$

However, it is pointed out in ref. 2 that for damping energy ratios typical of rubber the exact relation:

$$\psi = (1 - e^{-2\pi/M})$$

is preferable. This leads to values of M shown in Table 8.3, which correspond to the values of ψ given in Table 8.2.

It should be noted that when several components, e.g. two identical rubber couplings, are used in series the dynamic magnifier of the combination is given by:

$$\left(\frac{1}{M}\right)^2 = \left(\frac{1}{M_1}\right)^2 + \left(\frac{1}{M_2}\right)^2 + \left(\frac{1}{M_3}\right)^2 + \cdots \tag{13}$$

(this is an empirical rule, recommended in ref. 5)

An example of drive shaft design

The application of these principles is best illustrated by a worked example. Fig. 8.1 represents an engine coupled by way of twin multiple-bush type rubber couplings and an intermediate steel shaft to an eddy current dynamometer, with dynamometer starting.
Engine specification is as follows:

4 cylinder 4-stroke gasolene engine
Swept volume 2.0 litre, bore 86 mm, stroke 86 mm

Maximum torque	110 Nm at 4000 rev/min
Maximum speed	6000 rev/min
Maximum power output	65 kW
Maximum bmep	10.5 bar
Moments of inertia	$I_e = 0.25\,\text{kgm}^2$
	$I_b = 0.30\,\text{kgm}^2$

Table 8.4 indicates a service factor of 4.8, giving a design torque of $110 \times 4.8 = 530$ Nm.

It is proposed to connect the two couplings by a steel shaft of the following dimensions:

Diameter	$D = 40\,\text{mm}$
Length	$L = 500\,\text{mm}$
Modulus of rigidity	$G = 80 \times 10^9\,\text{Pa}$

From eq. (9a), torsional stress $\tau = 42$ Mpa, very conservative.
From eq. (10a)

$$C_s = \frac{\pi \times 0.04^4 \times 80 \times 10^9}{32 \times 0.5} = 40\,200\;\text{N m/rad}$$

Table 8.4. *Service factors*

	Number of cylinders									
	Diesel					*Gasolene*				
Dynamometer type	1/2	3/4/5	6	8	10+	1/2	3/4/5	6	8	10+
Hydraulic	4.5	4.0	3.7	3.3	3.0	3.7	3.3	3.0	2.7	2.4
Hyd.+dyno. start	6.0	5.0	4.3	3.7	3.0	5.2	4.3	3.6	3.1	2.4
Eddy current (EC)	5.0	4.5	4.0	3.5	3.0	4.2	3.8	3.3	2.9	2.4
EC+dyno. start	6.5	5.5	4.5	4.0	3.0	5.7	4.8	3.8	3.4	2.4
d.c.+dyno. start	8.0	6.5	5.0	4.0	3.0	7.2	5.8	4.3	3.4	2.4

Consider first the case when rigid couplings are employed:

$$n_c = \frac{60}{2\pi} \sqrt{\frac{40\,200 \times 0.55}{0.25 \times 0.30}} = 5185 \text{ c.p.m.}$$

For a four cylinder 4-stroke we have seen that the first major critical occurs at order $N_o = 2$, corresponding to an engine speed of 2592 rev/min. This falls right in the middle of the engine speed range and is clearly unacceptable. This is a typical result to be expected if an attempt is made to dispense with flexible couplings.

The resonant speed needs to be reduced and it is a common practice to arrange for this to lie between either the cranking and idling speeds or between the idling and minimum full load speeds. In the present case these latter speeds are 500 rev/min and 1000 rev/min respectively. This suggests a critical speed N_c of 750 rev/min and a corresponding resonant frequency $n_c = 1500$ cycles /min.

This calls for a reduction in the torsional stiffness in the ratio:

$$\left(\frac{1500}{5185}\right)^2$$

i.e. to 3364 N m/rad.

The combined torsional stiffness of several elements in series is given by:

$$\frac{1}{C} = \frac{1}{C_1} + \frac{1}{C_2} + \frac{1}{C_3} + \cdots \tag{14}$$

This equation indicates that the desired stiffness could be achieved by the use of two flexible couplings each of stiffness 7480 N m/rad. A manufacturer's catalogue shows a multi-bush coupling having the following characteristics:

Maximum torque	814 N m (adequate)
Rated torque	170 N m
Maximum continuous vibratory torque	±136 N m
Shore (IHRD) hardness	50/55
Dynamic torsional stiffness	8400 N m/rad

Substituting this value in eq. (14) indicates a combined stiffness of 3800 N m/rad. Substituting in eq. (1) gives $n_c = 1573$, corresponding to an engine speed of 786 rev/min, which is acceptable.

It remains to check on the probable amplitude of any torsional oscillation at the critical speed. Under no-load conditions the imep of the engine is likely to be in the region of 2 bar, indicating, from eq. (5a), a mean turning moment $M_{mean} = 8$ N m. From Table 8.1, p factor = 1.91, giving $T_{ex} = 15$ N m per cylinder, $\sum T_{ex} = 4 \times 15 = 60$ N m.

Table 8.3 indicates a dynamic magnifier $M = 10.5$, combined dynamic magnifier from eq. (13) = 7.4.

The corresponding value of the vibratory torque, from eq. (7), is then:

$$T_v = \frac{60 \times 7.4}{(1 + 0.25/0.30)} = \pm 242 \text{ N m}$$

This is in fact outside the coupling continuous rating of ± 136 N m but multiple bush couplings are tolerant of brief periods of quite severe overload and this solution should be acceptable provided the engine is run fairly quickly through the critical speed. An alternative would be to choose a coupling of similar stiffness using SBR bushes of 60/65 hardness. Table 8.3 shows that the dynamic magnifier is reduced from 10.5 to 2.7, with a corresponding reduction in T_v.

If in place of an eddy current dynamometer we were to employ a d.c. machine the inertia I_b would be of the order of 1 kgm^2, 4 times greater.

This has two adverse effects:

- service factor, from Table 8.4, increased from 4.8 to 5.8
- the denominator in eq. (7) is reduced from $(1 + 0.25/0.30\ ) = 1.83$ to $(1 + 0.25/1.0) = 1.25$, corresponding to an increase in the vibratory torque for a given exciting torque of nearly 50%.

This is a general rule: the greater the inertia of the dynamometer the more severe the torsional stresses generated by a given exciting torque.

An application of eq. (1) shows that for the same critical frequency the combined stiffness must be increased from 3364 N m/rad to 5400 N m/rad. We can meet this requirement by changing the bushes from Shore Hardness 50/55 to Shore Hardness 60/65, increasing the dynamic torsional stiffness of each coupling from 8400 N m/rad to 14 000 N m/rad. (in general the usual range of Hardness Numbers, from 50/55 to 75/80, corresponds to a stiffness range of about 3:1, a useful degree of flexibility for the designer).

Eq. (1) shows that with this revised coupling stiffness n_c changes from 1573 cycles/min to 1614 cycles/min, and this should be acceptable. The oscillatory torque generated at the critical speed is increased by the two factors mentioned above, but reduced to some extent by the lower dynamic magnifier for the harder rubber, $M = 8.6$ against $M = 10.5$. As before, prolonged running at the critical speed should be avoided.

For completeness, we should check the whirling speed from eq. (11) . The mass of the shaft per unit length is: $W_s = 9.80$ kg/m.

$$N_w = \frac{30\pi}{0.50^2} \sqrt{\frac{200 \times 10^9 \times \pi \times 0.044^4}{64 \times 9.80}} = 19\ 100 \text{ rev/min}$$

The mass of the shaft + half couplings is found to be 12 kg and the combined radial stiffness 33.6 MN/m. From eq (12a):

$$N_t = \frac{30}{\pi} \sqrt{\frac{33.6 \times 10^6}{12}} = 16\ 000 \text{ rev/min}$$

then from eq. (12b), whirling speed $= 12\ 300$ rev/min, satisfactory

Note however that, if shaft length were increased from 500 mm to 750 mm,

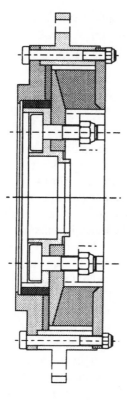

Figure 8.10. *Annular type rubber coupling*

whirling speed would be reduced to about 7300 rev/min, barely acceptable. This is a common problem, usually dealt with by the use of tubular shafts, which have much greater transverse stiffness for a given mass.

There is no safe alternative, when confronted with an engine of which the characteristics differ significantly from any run previously on a given testbed, to following through this design procedure.

An alternative solution

The above worked example makes use of two multiple-bush type rubber couplings with a solid intermediate shaft. An alternative is to make use of a conventional propeller shaft with two universal joints, as used in road vehicles, with the addition of a coupling incorporating an annular rubber element in shear to give the necessary torsional flexibility. These couplings, Fig. 8.10, are generally softer than the multiple bush type for a given torque capacity, but are less tolerant of operation near a critical speed. If it is decided to use a conventional universal joint shaft, the supplier should be informed of the maximum speed at which it is to run. This will probably be much higher than is usual in the vehicle and may call for tighter than usual limits on balance of the shaft.

Shock loading of couplings due to cranking, irregular running and torque reversal

Systems for starting and cranking engines are described in Chapter 7, p.130, where it is emphasized that during engine starting severe transient torques can arise. These have been known to result in the failure of flexible couplings of apparently adequate torque capacity. The maximum torque that can be necessary to get the engine over tdc, or that can be generated at first firing, should be estimated and checked against maximum coupling capacity.

Irregular running or imbalance between the power generated by individual cylinders can give rise to exciting torque harmonics of lower order than expected in a multi-cylinder engine and should be borne in mind as a possible source of rough running. Finally there is the possibility of momentary torque reversal when the engine comes to rest on shut-down.

However, the most serious problems associated with the starting process arise when the engine first fires. Particularly when, as is common practice, the engine is motored to prime the fuel injection pump, the first firing impulses can give rise to severe shocks. Annular type rubber couplings, Fig. 8.10, can fail by shearing under these conditions: the authors have known cases when it was necessary to fit a torque limiter or slipping clutch to deal with this problem.

Selection of coupling torque capacity

Initial selection is based on the maximum rated torque with consideration given to the type of engine and number of cylinders, dynamometer character- istics and inertia. Table 8.4, reproduced by courtesy of Twiflex Ltd, shows recommended service factors for a range of engine and dynamometer combina- tions. The rated torque multiplied by the service factor must not exceed the permitted maximum torque of the coupling.

Other manufacturers may adopt different rating conventions, but Table 8.4 gives valuable guidance as to the degree of severity of operation for different situations. Thus, for example, a single cylinder diesel engine coupled to a d.c. machine with dynamometer start calls for a margin of capacity three times as great as an 8-cylinder gasolene engine coupled to a hydraulic dynamometer.

Fig. 8.11 shows the approximate range of torsional stiffness associated with three types of flexible coupling: the annular type as illustrated in Fig. 8.10, the multiple bush design of Fig 8.8, and a development of the multiple bush design which permits a greater degree of misalignment and makes use of double-ended

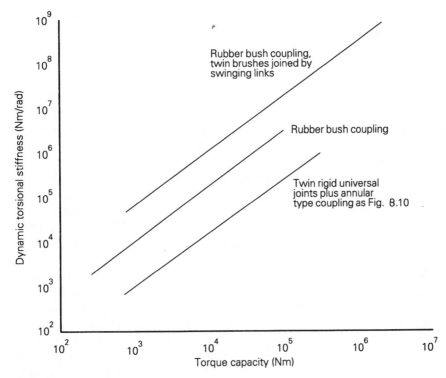

Figure 8.11. *Ranges of torsional stiffness for different types of rubber coupling*

bushed links between the two halves of the coupling. The stiffnesses of Fig. 8.11 refer to a single coupling.

The role of the engine clutch

Vehicle engines are invariably fitted with a clutch and this may or may not be retained on the test bed. The advantage of retaining the clutch is that it acts as a torque limiter under shock or torsional vibration conditions. The disadvantages are that it may creep, particularly when torsional vibration is present, leading to ambiguities in power measurement, while it is usually necessary, when the clutch is retained, to provide an outboard bearing. Clutch disc springs may have limited life under test bed conditions.

Guarding of coupling shafts

The failure of a high speed coupling shaft can be extremely dangerous, as very large amounts of energy can be released. A really substantial guard, preferably a steel tube not less than 6 mm thick, split and flanged in the horizontal plane, is an essential precaution.

Balancing of drive line components

This is a matter which, in the authors' experience, is often not taken sufficiently seriously and can lead to a range of troubles, including damage (which can be very puzzling) to bearings, unsatisfactory performance of such items as torque transducers, transmitted vibration to unexpected locations, and serious drive line failures. Particular care should be taken in the choice of couplings for torque shaft dynamometers: couplings such as the multiple disc type, Fig. 8.7, cannot be relied upon to centre these devices sufficiently accurately.

Conventional universal joint type cardan shafts are often required to run at higher speeds in test bed applications than is usual in vehicles; when ordering, the maximum speed should be specified and, possibly, a more precise level of balancing than standard specified.

BS 5265[9], *Mechanical balancing of rotating bodies*, gives a valuable discussion of the subject and specifies 11 Balance Quality Grades. Drive line components should generally be balanced to Grade G 6.3, or, for very high speeds, to grade G 2.5. The Standard gives a logarithmic plot of the permissible displacement of the centre of gravity of the rotating assembly from the geometrical axis against speed. To give an idea of the magnitudes involved, G 6.3 corresponds to a displacement of 0.06 mm at 1000 rev/min, falling to 0.01 mm at 5000 rev/min.

Alignment of engine and dynamometer

This is a fairly complex and quite important matter. For a full treatment and description of alignment techniques see ref. 4. Temperature effects, and the movement of the engine on its flexible mountings when on load, should be taken into account and if possible the mean position should be determined. The laser-based alignment systems now available greatly reduce the effort and skill required to achieve satisfactory levels of accuracy. In particular, they are able to bridge large gaps between flanges without any compensation for droop and deflection of arms carrying dial indicators, a considerable problem with conventional alignment methods.

There are essentially three types of alignment to be considered:

- rubber bush and flexible disc couplings should be aligned as perfectly as possible, as any misalignment encourages heating of the elements and fatigue damage
- gear type couplings require a minimum misalignment of about 0.04° to encourage the maintenance of an adequate lubricant film between the teeth
- most manufacturers of universal joint propeller shafts recommend a small

degree of misalignment to prevent brinelling of the universal joint needle rollers. Note that it is essential, in order to avoid induced torsional oscillations, that the two yokes of the intermediate shaft joints should lie in the same plane.

Distance between end flanges can be critical, as incorrect positioning can lead to the imposition of axial loads on bearings of engine or dynamometer.

Engine to dynamometer coupling: summary of design procedure

(1) Establish speed range and torque characteristic of engine to be tested. Is it proposed to run the engine on load throughout this range?
(2) Make a preliminary selection of a suitable drive shaft. Check that maximum permitted speed is adequate. Check drive shaft stresses and specify material. Look into possible stress raisers.
(3) Check manufacturer's recommendations regarding load factor and other limitations.
(4) Establish rotational inertias of engine and dynamometer and stiffness of proposed shaft and coupling assembly. Make a preliminary calculation of torsional critical speed from eq. (1). (in the case of large multi-cylinder engines consider making a complete analysis of torsional behaviour)
(5) Modify specification of shaft components as necessary to position torsional criticals suitably. If necessary consider use of viscous torsional dampers.
(6) Calculate vibratory torques at critical speeds and check against capacity of shaft components. If necessary specify 'no go' areas for speed and load.
(7) Check whirling speeds.
(8) Specify alignment requirements.
(9) Design shaft guards.

Notation

frequency of torsional oscillation	n cycles/min
critical frequency of torsional oscillation	n_c cycles/min
stiffness of coupling shaft	C_s N m/rad
rotational inertia of engine	I_e kg m^2
rotational inertia of dynamometer	I_b kg m^2
amplitude of exciting torque	T_{ex} N m
amplitude of torsional oscillation	θ rad
static deflection of shaft	θ_0 rad
dynamic magnifier	M
dynamic magnifier at critical frequency	M_c
order of harmonic component	N_o
number of cylinders	N_{cyl}

mean turning moment	M_{mean} N m
indicated mean effective pressure	p_i bar
cylinder bore	B mm
stroke	S mm
component of tangential effort	T_m N m
amplitude of vibratory torque	T_v N m
engine speed corresponding to n_c	N_c rev/min
maximum shear stress in shaft	τ N/m^2
whirling speed of shaft	N_w rev/min
transverse critical frequency	N_t cycles/min
dynamic torsional stiffness of coupling	C_c N m/rad
damping energy ratio	ψ
modulus of elasticity of shaft material	E Pa
modulus of rigidity of shaft material	G Pa

(for steel, $E = 200 \times 10^9$ Pa, $G = 80 \times 10^9$ Pa)

References

1. Den Hartog, J.P. (1956) *Mechanical Vibrations*, McGraw Hill, Maidenhead, UK.
2. Ker Wilson, W. (1963) *Practical Solution to Torsional Vibration Problems* (5 Vols), Chapman and Hall, London.
3. Ker Wilson, W. (1959) *Vibration Engineering*, Griffin, London.
4. Neale, M.J. *et al.* (1998) *Couplings and Shaft Alignment*, Mechanical Engineering Publications, London.
5. *Rulebook*, Chapter 8. Shaft vibration and alignment, Lloyd's Register of Shipping, London.
6. Pilkey, W.D. (1997) *Peterson's Stress Concentration Factors*, Wiley, New York.
7. Young, W.C. (1989) *Roark's Formulas for Stress and Strain*, McGraw Hill, New York.
8. BS 6613 *Methods for Specifying Characteristics of Resilient Shaft Couplings.*
9. BS 5265 Parts 2 and 3 *Mechanical Balancing of Rotating Bodies.*

Further reading

Nestorides, E.J. (1958) *A Handbook of Torsional Vibration*, Cambridge University Press, Cambridge.
BS 4675 Parts 1 and 2 *Mechanical Vibration in Rotating Machinery.*
BS 6861 Part 1 *Method for Determination of Permissible Residual Unbalance.*
BS 6716 *Guide to Properties and Types of Rubber.*

9 Measurement of torque, power, speed and fuel consumption; acceptance and type tests, accuracy of the measurements

This chapter deals with the special problems involved in the *accurate* measurement of those (measurable) characteristics of an engine that are of particular interest and concern to a customer. These are likely to be its power output, fuel consumption and, possibly, its lubricant consumption. It is a subject of great practical importance, since it is an area in which disputes between manufacturer and customer are most likely to arise.

It is also an area in which the achievement of a high level of accuracy is much more difficult than is commonly appreciated, even by engineers, and much of the chapter is devoted to a detailed discussion of the problems involved.

Key factors and relationships

There are only a few factors involved:

Measured:
Dynamometer torque arm radius	R m
Torque arm transducer force	F N
Engine speed	N rev/min
Fuel consumption rate, by volume or mass	$\dot{V}$ l/h
	$\dot{m}$ kg/h
Fuel density	ρ kg/l
Lubricant consumption rate	M_o kg/100hr

Derived:
Engine torque	T Nm
Engine power output	P kW
Specific fuel consumption	sfc kg/kWh

There are a few simple relationships between these quantities:

$$\dot{m} = \rho \dot{V}$$

$$T = FR$$

$$P = 2\pi \frac{N}{60} T \times 10^{-3}$$

$$sfc = \frac{\dot{m}}{P}$$

Note that the power P is the product of three measurements, F, R, and N while the specific fuel consumption is the product of four or five, F, R, N, $\dot{m}$, and possibly ρ.

It follows that the accuracy of these measurements is less than the accuracy of each component measurement, see Chapter 20 for a formal treatment of this subject.

Measurement of torque: trunnion-mounted machines

The great majority of dynamometers use this method of torque measurement, the essential feature of which is that the power absorbing (or power producing) element of the machine is mounted on bearings coaxial with the machine shaft and the torque is measured by some kind of transducer acting tangentially at a known radius from the machine axis. In traditional machines the transducer consisted of a combination of dead weights and spring balance, Fig. 9.1(a). As the stiffness of the balance was limited it was necessary to adjust its position depending on the torque, to ensure that the force measured was accurately tangential.

Modern machines (Fig. 9.1(b)) use a transducer, almost invariably of the strain gauge type, together with an appropriate bridge circuit and amplifier. The strain gauge transducer has the advantage of being extremely stiff, so that no positional adjustment is necessary, but the disadvantage of a finite fatigue life; there is a possibility of failure after a (very large) number of load applications.

The machine is usually mounted on trunnion bearings, typically a combination of a ball bearing (for axial location) and a roller bearing. These bearings operate under unfavourable conditions, with no perceptible angular movement, and are consequently prone to brinelling, or local indentation of the races, and to fretting. This is aggravated by vibration that may be transmitted from the engine, and periodical inspection is desirable. The Schenck dynamometer, Fig. 9.1(c), replaces the trunnion bearings by two radial flexures, thus eliminating possible friction and wear, but at the expense of the introduction of torsional stiffness, of reduced capacity to withstand axial loads and of possible ambiguity regarding the true centre of rotation, particularly under side loading.

For extreme accuracy, such as is required for instance in national standards

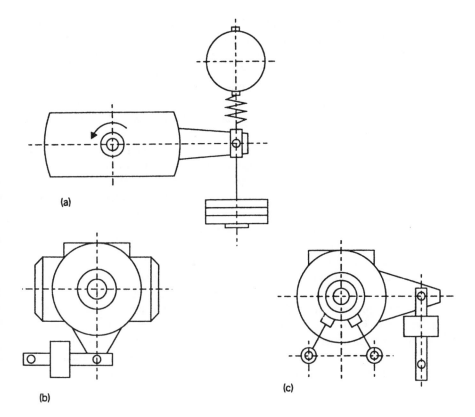

Figure 9.1. *Dynamometer mounting and torque measurement: (a) dead weights and spring balance; (b) torque transducer; (c) Schenck flexure mounting*

laboratories, the trunnion bearings are sometimes arranged with driven rotating outer races. The two bearings are driven in opposite directions and all parasitic friction should be eliminated.

Calibration and the assessment of errors in torque measurement

We have seen that in a conventional dynamometer torque T is measured as a product of torque arm radius R and transducer force F. Calibration is invariably by means of a *calibration arm*, supplied by the manufacturer, which is bolted to the dynamometer carcase and carries dead weights which apply a load at a certified radius. The manufacturer certifies the distance between the axis of the weight hanger bearing and an axis defined by a line joining the centres of the trunnion bearings (not the axis of the dynamometer, which indeed need not precisely coincide with the axis of the trunnions).

There is no way, apart from building an elaborate fixture, in which the dynamometer user can check the accuracy of this dimension: he is entirely in the hands of the manufacturer. The arm should be stamped with its effective length. For *R* and *D* machines of high accuracy the arm should be stamped for the specific machine.

The 'dead weights' will in fact be more correctly termed 'standard masses'. They should be certified by an appropriate Standards Authority. However, the force they exert on the calibration arm is the product of their mass and the local value of '*g*'. This is usually assumed to be 9.81 m/s^2 and constant : in fact this value is only correct at sea level and a latitude of about $47°$[1]. It increases towards the poles and falls towards the equator, with local variations. As an example, a machine calibrated in London, where $g = 9.81 \text{ m/s}^2$, will read 0.13% high if recalibrated in Sydney, Australia and 0.09% low if recalibrated in St. Petersburg without correcting for the different local values of *g*. These are not negligible variations if one is hoping for accuracies better than 1%.

The actual process of calibrating a dynamometer with dead weights, if treated rigorously, is not entirely straightforward. We are confronted with the facts that no transducer is perfectly linear in its response, and no linkage is perfectly frictionless. We are then faced with the problem of adjusting the system so as to ensure that the (inevitable) errors are at a minimum throughout the range.

A suitable calibration procedure for a machine using a typical strain-gauge load cell for torque measurement is as follows. The dynamometer should not be coupled to the engine. After the system has been energized long enough to warm up the load cell output is zeroed with the machine in its normal no-load running condition (cooling water on, etc.). Dead weights are then added to produce approximately the rated maximum torque of the machine. This torque is calculated and the digital indicator set to this value.

The weights are removed, the zero reading noted, and weights are added, preferably in 10 equal increments, the cell readings being noted. The weights are removed in reverse order and the readings again noted.

The procedure described above means that the load cell indicator was set to read zero *before* any load was applied (it did not necessarily read zero after the weights had been added and removed), while it was adjusted to read the *correct* maximum torque when the appropriate weights had been added.

We now ask: is this setting of the load cell indicator the one that will minimize errors throughout the range and are the results within the limits of accuracy claimed by the manufacturer?

Let us assume we apply this procedure to a machine having a nominal rating of 600 Nm torque, and that we have 6 equal weights, each calculated to impose a torque of 100 Nm on the calibration arm. Table 9.1 shows the indicated

Table 9.1.   *Dynamometer calibration*

Mass (kg)	Applied torque (Nm)	Reading (Nm)	Error (Nm)	Error (% reading)	Error (% full scale)
0	0	0.0	0.0	0.0	0.0
10	100	99.5	−0.5	−0.5	−0.083
30	300	299.0	−1.0	−0.33	−0.167
50	500	500.0	0.0	0.0	0.0
60	600	600.0	0.0	0.0	0.0
40	400	400.5	+0.5	+0.125	+0.083
20	200	200.0	0.0	0.0	0.0
0	0	0.0	0.0	0.0	0.0

torque readings for both increasing and decreasing loads, together with the calculated torques applied by the weights. The corresponding *errors*, or the differences between torque applied by the calibration weights and the indicated torque readings are plotted in Fig. 9.2. The machine is claimed to be accurate to within ± 0.25% of nominal rating and these limits are shown. It will be clear that the machine meets the claimed limits of accuracy and may be regarded as satisfactorily calibrated.

It is usually assumed, though it is not necessarily the case, that hysteresis effects, manifested as differences between observed torque with rising load and with falling load, are eliminated when the machine is running, due to vibration, and it is a common practice when calibrating to knock the machine carcase lightly with a soft mallet after each load change to achieve the same result.

It is certainly *not* wise to assume that the ball joints invariably used in calibration arm and torque transducer links are frictionless. These bearings are designed for working pressures on the projected area of the contact in the range 15 to 20 MN/m^2 and a 'stick slip' coefficient of friction at the ball surface of, at a minimum, 0.1, is to be expected. This clearly affects the effective arm length (in either direction) and must be relaxed by vibration.[3]

Some large dynamometers are fitted with torque multiplication levers, reducing the size of the calibration masses and, for large machines, torque calibration by way of load cells or proving rings is sometimes practised, involving a more complex 'audit trail' to refer the calibration back to national standards.

It is important, when calibrating an eddy-current machine, that the water pressure in the casing should be at operational level, since pressure in the transfer pipes can give rise to a parasitic torque. Similarly any disturbance to

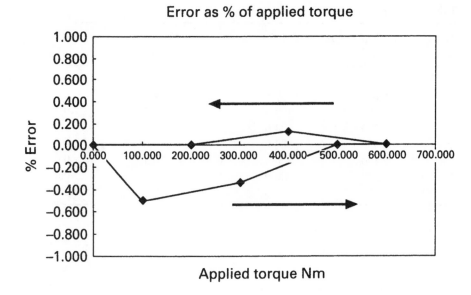

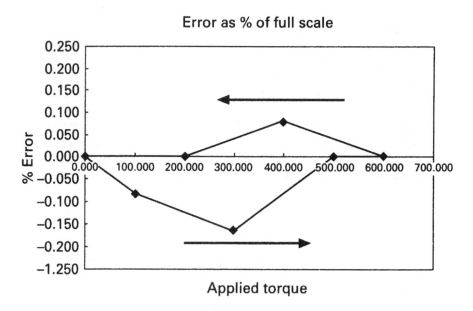

Figure 9.2.   Dynamometer calibration curves

the run of electrical cables to the machine must be avoided once calibration is completed. Finally, it is possible, particularly with electrical dynamometers with forced cooling, to develop small parasitic torques due to air discharged non-radially from the casing. It is an easy matter to check this by running the machine uncoupled under its own power and noting any change in torque.

Experience shows that a high grade dynamometer such as would be used for research work, after careful calibration, may be expected to give a torque indication that does not differ from the absolute value by more than about $\pm 0.25\%$ of the full load torque rating of the machine. Systematic errors such as inaccuracy of torque arm length or wrong assumptions regarding the value of g will certainly diminish as the torque is reduced but other errors will be little affected: it is safer to assume a band of uncertainty of constant width, this implies, for example, that a machine rated at 400 N m torque with an accuracy of ± 0.25 % will have an error band of ± 1 N. At 10% of rated torque, this implies that the true value may lie between 39 and 41 N m. It is well to match the size of the dynamometer as closely as possible with the rating of the engine.

All load cells used by reputable dynamometer manufacturers will compensate for changes in temperature, though their rate of response to a change may vary. They will not however be able to compensate for internal temperature gradients induced, for example, by air blasts from ventilation fans or radiant heat from exhaust pipes.

The subject of calibration and accuracy of dynamometer torque measurement has been dealt with in some detail, but this is probably the most critical measurement that the test engineer is called upon to make, and one for which a high standard of accuracy is expected but not easily achieved. Calibration and certification of the dynamometer and its associated system should be carried out at the very least once a year, and following any system change or major component replacement.

Torque measurement under accelerating and decelerating conditions

With the increasing interest in transient testing it is essential to be aware of the effect of speed changes on the 'apparent' torque measured by a trunnion-mounted machine.

The basic principle is simple:

Inertia of dynamometer rotor	I kg m^2
Rate of increase in speed	$\dot{\omega}$ rad/s^2
	$\dot{N}$ rpm/s
Input torque to dynamometer	T_1 N m
Torque registered by dynamometer	T_2 N m

$$T_1 - T_2 = I\dot{\omega} = \frac{2\pi \dot{N}I}{60} \, \text{Nm}$$

$$= 0.1047 \, \dot{N}I \, \text{Nm}$$

To illustrate the significance of this correction, a typical eddy current dynamometer capable of absorbing 150 kW with a maximum torque of 500 Nm has a rotor inertia of 0.11 kg m^2. A d.c. regenerative machine of equivalent rating has an inertia of 0.60 kg m^2.

If these machines are coupled to an engine that is accelerating at the comparatively slow rate of 100 rpm/s the first machine will read the torque low by an amount:

$$T_1 - T_2 = 0.1047 \times 100 \times 0.11 = 1.15 \, \text{Nm}$$

while the second will read low by 6.3 N m.

If the engine is decelerating the machines will read high by the equivalent amount. With computer processing of the data corrections can be made with appropriate software.

Much larger rates of speed change are demanded in some transient test sequences, see Chapter 15, and this can represent a serious limitation, particularly in the use of d.c. dynamometers.

Measurement of power by torque shaft dynamometer

A torque shaft dynamometer is mounted in the drive shaft between engine and brake. It consists essentially of a flanged torque shaft fitted with strain gauges, and designs are available both with slip rings and with signal transmission by short wave radio. Figure 9.3 shows a typical slip ring type torque transducer

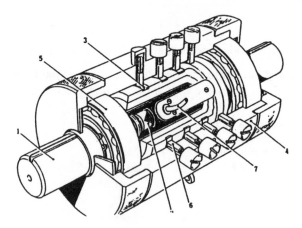

Figure 9.3. *Slipring torque shaft dynamometer*

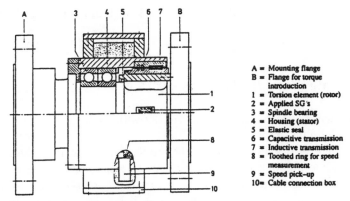

A = Mounting flange
B = Flange for torque
    introduction
1 = Torsion element (rotor)
2 = Applied SG's
3 = Spindle bearing
4 = Housing (stator)
5 = Elastic seal
6 = Capacitive transmission
7 = Inductive transmission
8 = Toothed ring for speed
    measurement
9 = Speed pick–up
10= Cable connection box

Figure 9.4.   *Brushless torque shaft dynamometer*

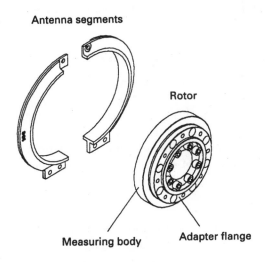

Figure 9.5.   *Disc type shaft dynamometer*

intended to be supported by the shafts while Fig. 9.4 is a brushless unit intended for rigid mounting.

The main advantage of the transmission dynamometer is that it avoids the necessity, discussed above, of applying torque corrections under transient conditions. It does, however, require very careful installation to avoid the imposition of bending or axial stresses on the torsion sensing element. It is thus more easily used with large rigidly mounted engines than with smaller engines carried on flexible mountings.

A recent development is the 'disc' type transmission dynamometer, Fig. 9.5, which may be bolted directly to the brake flange and considerably simplifies installation.

Measurement of speed and power

Measurement of speed using a shaft encoder with analogue or digital display is in principle quite simple.

Measurement of power, which is the product of torque and speed, raises the important question of sampling time. Engines never run totally steadily and the torque transducer and speed signals invariably fluctuate. An instantaneous snap reading will not necessarily, or even probably, be identical with a longer-term average. Choice of sampling time and of the number of samples to be averaged is a matter of compromise. Under transient conditions there may be no choice but to take snap readings. This question is discussed in more detail below.

Measurement of fuel consumption

This presents a number of problems which may be solved in several ways, depending on the purpose of the test. For a really accurate result there is no substitute for a cumulative measurement under steady running conditions. This also deals with the problem of short-term fluctuations in torque and speed.

Cumulative flowmeters

There are several types of fuel gauge intended for cumulative measurement on the market. Figure 9.6 shows a volumetric gauge in which an optical system gives a precise time signal at the start and end of one of a choice of calibrated volumes. This signal actuates a counter, giving a precise value for the number of engine revolutions made during the consumption of the measured volume of fuel. At the same time a computer is set to calculate an average dynamometer torque value during the interval. In the experience of the authors, the consumption period should not be less than about 30 s.

Figure 9.7 shows a gauge intended to meter a mass rather than a volume of fuel, consisting essentially of a vessel mounted on a weighing machine from which fuel is drawn by the engine. The signals are processed in the same way as for the volumetric gauge. Figure 9.8 shows a further type of gravimetric fuel gauge, in which a cylindrical float is suspended from a force transducer in a cylindrical vessel. The change in flotation force is then directly proportional to the change of mass of fuel in the vessel. This design of gauge has the particular

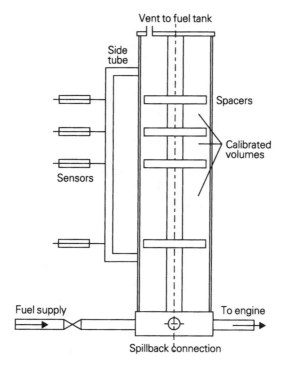

Figure 9.6. *Volumetric fuel consumption gauge*

advantage that it is insensitive to vertical accelerations and is thus suitable for use on shipboard.

These gravimetric gauges have the advantage over the volumetric type that the metered mass may be chosen at will, and a common measuring period may be chosen, independent of fuel consumption rate. The specific fuel consumption is derived directly from only three measured quantities: the mass of fuel consumed, the number of engine revolutions during the consumption of this mass and the mean torque. This is another reason for the inherently greater accuracy of cumulative fuel consumption measurement; rate measurements involve four or, in the case of volumetric meters, five measured quantities.

All gauges of this type have to deal with the problem of fuel spillback from engines fitted with fuel injection systems, and Fig. 9.9 shows a circuit incorporating a gravimetric fuel gauge which deals with this matter. When a fuel consumption measurement is to be made a solenoid valve diverts the spillback flow, normally returned to the header tank during an engine test, into the bottom of the fuel gauge. (It is not satisfactory to return the spillback to a fuel filter downstream of the fuel gauge, since the air and vapour always

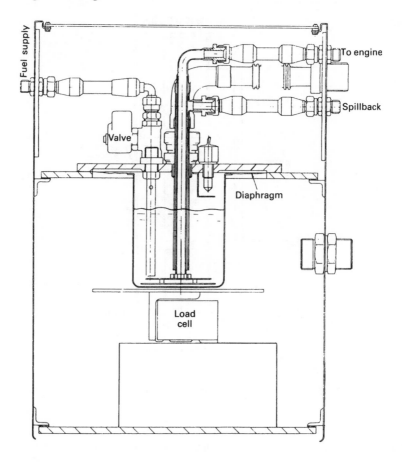

Figure 9.7.   *Direct weighing fuel gauge*

present in the spilled fuel lead to variations in the fuel volume between fuel gauge and engine, and thus to incorrect values of fuel consumption.)

Rate meters

There are many different designs of rate meter on the market and the choice of the most suitable unit for a given application is not easy. The following factors need to be taken into account:

- volumetric or gravimetric
- absolute level of accuracy
- turn-down ratio
- sensitivity to temperature and fuel viscosity

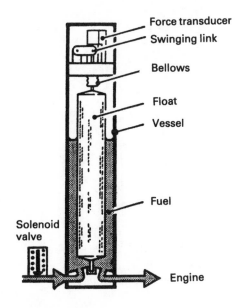

Figure 9.8. *Gravimetric fuel gauge*

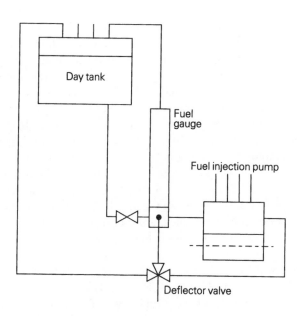

Figure 9.9. *Fuel measurement with spillback*

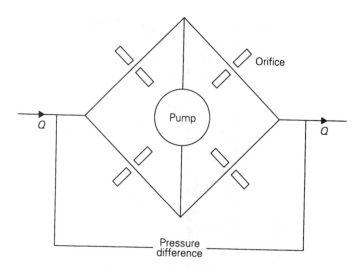

Figure 9.10. *Principle of Flowtron flowmeter*

- pressure difference required to operate
- wear resistance and tolerance of dirt
- analogue or impulse-counting readout
- suitability for stationary/in-vehicle use.

Several designs of positive-displacement (volumetric) fuel gauges make use of a four-piston metering unit with the cylinders arranged radially around a single-throw crankshaft. Crankshaft rotation is transmitted magnetically to a pulse output flow transmitter. Cumulative flow quantity and instantaneous flow rate are indicated and these meters are suitable for in-vehicle use. A high turn-down ratio is claimed and one design includes a pressure-sensing system to eliminate leakage errors. A disadvantage is the appreciable pressure drop, which may approach 1 bar, required to drive the metering unit. For large flow rates metering units employing meshing helical rotors are usual.

The well-known 'Flowtron' meter makes use of the principle of the 'hydraulic Wheatstone bridge', Fig. 9.10. It is easily shown that this device gives rise to a pressure-difference signal that is directly proportional to mass flow rate. The device is however inevitably sensitive to fuel viscosity, since the coefficient of discharge of the measuring orifices is a function of Reynolds Number.

Other mass flowmeters make use of the Coriolis effect, in which the fuel is passed through a vibrating U-tube, Fig. 9.11. These flowmeters, suitable for larger engines, are claimed to have a high level of accuracy and repeatability, although care is required in their installation to avoid the effects of external vibration.

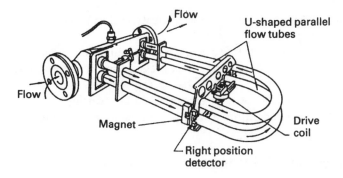

Figure 9.11.   *Coriolis effect flowmeter*

Where very high accuracy is not required fuel gauges operating on the 'Rotameter' principle, in which a conical float moves in a tapered vertical tube, may be considered, though only for stationary applications.

A disadvantage of the ratemeter method of measurement is that, where spillback is involved, it is necessary to use two sensing units, one in the flow line and one in the spill line. This increases cost and complexity and reduces accuracy.

Where a fuel consumption system combining the accuracy of a cumulative measurement with the ability of a rate meter to deal with transient conditions is required there is no substitute for the use of a rate meter in series with a cumulative meter. The cumulative device can then act as a built-in calibration system for the rate meter.

Fuel consumption measurements: gaseous fuels

For these fuels consumption measurements are made by gas flow meters. Metering of gases is a more difficult matter than liquid metering. The density of a gas is sensitive to both pressure and temperature, both of which must be known when, as is usually the case, the flow is measured by volume. Also the pressure difference available to operate the meter is limited. The traditional domestic gas meter contains four chambers separated by bellows and controlled by slide valves. Successive increments in volume metered are quite large, so that instantaneous or short-term measurements are not possible.

Other types, capable of indicating smaller increments of flow, make use of rotors having sliding vanes or meshing rotors. See for example Fig. 10.8. (p.197)

In the case of natural gas supplied from the mains flow measurement is a fairly simple matter. Pressure and temperature at the meter must be measured, and accurate data on gas properties (density, calorific value, etc.) will be available from the supplier.

Measurement of liquefied petroleum gas consumption is less straightforward, since this is stored as a liquid under pressure and is vaporized, reduced in pressure, and heated before reaching the engine cylinders. The gas meter must be installed in the line between the 'converter' and the carburetter. It is essential to measure the gas temperature at this point to achieve accurate results.

Measurement of lubricating oil consumption

This is one of the most difficult measurements associated with engine testing and, with the exception of total loss systems, such as are used for cylinder lubrication of large marine diesel engines, no entirely satisfactory method exists. The usual method is to adapt the engine for *dry sump* operation. The sump is arranged to drain to a separate receiver, the contents of which are monitored, either by weighing or by depth measurement.

Difficulties include:

- Oil consumption rate is very slow relative to the quantity in circulation. In a typical vehicle engine of lubricant capacity 6 litre the oil consumption rate will be in the region of zero to $25 \, cm^3/h$, or up to 0.5% of the volume in circulation.
- The quantity of oil adhering to the internal surfaces of the engine is very sensitive to temperature, as is the volume in transit to the receiver. This can give rise to large apparent variations in consumption rate.
- Apparent consumption is influenced by fuel dilution and by any loss of oil vapour.
- The rate of oil consumption tends to be very sensitive to conditions of load, speed and temperature. There is also a tendency to medium-term variations in apparent rate, due to such factors as 'ponding up' of oil in return drains and accumulation of air or vapour in the circuit.

For a critical analysis of the dry sump method see ref. 4, which also describes a statistical test procedure aimed at minimizing these random errors.

Choice of methods for measuring torque, power, speed and fuel consumption

(1) Many factors are involved in the choice of dynamometer, see Chapter 7. In the present connection consider particularly:

    (a) method of calibrating torque measurement

    (b) inertia of rotor and its influence on transient tests

(c) possible use of transmission dynamometer.

(2) Decide on appropriate data logging and computer instantaneous/averaged values of speed and torque.

(3) Decide on type of fuel consumption measurement: cumulative, rate or both. Volumetric or gravimetric.

(4) Design suitable fuel supply circuit to deal with any special features of engine fuel system.

The significance of the 'power curve': the effect of fuel properties

The effect of the condition (pressure, temperature, humidity and purity) of the combustion air on the mass rate of flow of oxygen in a form available for combustion that enters the engine is discussed in Chapter 10. In the case of a spark ignition engine the maximum power output is more or less directly proportional to this mass, since the air/fuel ratio at full throttle is closely controlled.

In the case of the diesel engine the position is different, and less clear cut. The maximum power output is generally determined by the maximum mass rate of flow delivered by the fuel injection pump operating at the maximum fuel stop position. However the setting of the fuel stop in fact determines the maximum volumetric fuel flow rate and this is one reason why fuel temperature should always be closely controlled in engine testing.

Fuel has a high coefficient of cubical expansion, lying within the range 0.001-0.002 per deg C (compared with water, for which the figure is 0.00021 per deg C). This implies that the mass of fuel delivered by a pump is likely to diminish by between 0.1% and 0.2% for each degree rise in temperature, a by no means negligible effect.

A further fuel property that is affected by temperature is its viscosity, significant because in general the higher the viscosity of the fuel the greater the volume that will be delivered by the pump at a given rack setting. This effect is likely to be specific to a particular pump design; as an example, one engine manufacturer regards a 'standard ' fuel as having a viscosity of 3 cSt at 40°C and applies a correction of $2\frac{1}{2}$% for each centiStoke departure from this base viscosity, the delivered volume increasing with increasing viscosity.

As a further factor, the specific gravity (density) of the fuel will also affect the mass delivered by a constant volume pump, though here the effect is obscured by the fact that, in general, the calorific value of a hydrocarbon fuel falls with increasing density, thus tending to cancel the change out.

Cell to cell correlation

It is usual for an engine manufacturer to wish to be reassured that all the test stands in his test department give the same answer and, in the experience of the authors, it is not unusual for the attempt to be made to answer this question by the apparently logical procedure of testing the *same* engine on all the beds. The result is invariably a disappointment, and can lead to expensive and unnecessary disputes between the manufacturer and the supplier of the test equipment.

It cannot be too strongly emphasized that an engine, however sophisticated its management system, is *not* suitable for duty as a standard source of torque. In Chapter 10 we discuss the very substantial changes in engine performance that can arise from changes in atmospheric (and hence in combustion air) conditions and, since we cannot run the engine on two testbeds at the same time, we are prone to these effects. In addition engine power output is sensitive to variations in fuel and in lubricating oil and cooling water temperature and it would be necessary to equalize these very carefully if meaningful results were to be hoped for.

Finally, it is unlikely that a set of test cells will be *totally* identical: apparently small differences in such factors as the layout of the ventilation air louvres and in the exhaust system can have a significant effect on performance.

A fairly good indication of the impossibility of using an engine as a standard in this way is contained in the Standard, BS 5514, discussed below. This Standard lists the 'Permissible deviation' in engine torque as measured repeatedly during a single test run on a single test bed as ±2%. This apparently wide tolerance is no doubt based on experience and, by implication, invalidates the use of an engine to correlate dynamometer performance.

There is no substitute for the careful and regular calibration of all the machines, following the procedure described above.

Acceptance and type tests, power test codes and correction factors

Relations with customers

Most of the work carried out in engine test installations is concerned with purely technical and engineering matters. Acceptance and type tests involve much more: they are concerned with the interface between the engineering and commercial worlds and between manufacturer and customer. The reputation of the manufacturer is directly involved and any ambiguity in the test procedures or in the interpretation of the results can lead to the loss of goodwill and even to litigation. This is particularly the case when the customer is a government department or one of the defence services.

It is therefore not surprising that tests of these kinds are the subject of extensive regulations, in many cases having statutory force, and not always easy to interpret. It is essential that the engineer or technician who finds himself concerned with this kind of work should make himself thoroughly familiar with these regulations; the more important English language documents are described briefly below.

Acceptance tests

These can range from the briefest acceptable running-in period plus a check on the general operation of the engine, carried out in accordance with the engine manufacturer's internal procedures and without the involvement of the customer, to elaborate and detailed tests carried out in the presence of the customer or his representative in order to ensure that the engine and its performance meet an agreed specification.

Type tests

Type tests are particularly associated with supplies to the Armed Forces, who invariably lay down the procedure in detail. A type test is designed to prove the whole performance of the engine under the conditions it may actually meet in service. It should cover everything necessary to prove beyond doubt or dispute that the required performance is met or exceeded. This means that a type test will involve prolonged running under conditions representative of service with the specified maintenance procedures and measurements of wear and of such factors as piston cleanliness.

It will be quite distinct from the relatively brief production tests which, while also checking that the required performance is achieved, are aimed primarily at ensuring that the quality of materials and workmanship and the production routine are up to the standard established by the type test.

Power test codes and correction factors

Several complex sets of rules are in general use for specifying the procedure for measuring the performance of an engine and for correcting this to standard conditions. The most significant for the English-speaking world are:

BS 5514[5]	Reciprocating internal combustion engines
ISO 3046[6]	Reciprocating internal combustion engines (identical to BS 5514)
BS AU 141[7]	Road vehicle diesel engines
Lloyd's Rule Book[8]	(primarily concerned with marine engines and installations)

(American) SAE Standards

| SAE J1995[9] | Engine power test code – spark ignition and compression ignition – gross power rating |
| SAE J1349[10] | Engine power test code – spark ignition and compression ignition – net power rating |

BS 5514/ISO 3046

Perhaps the best starting point for the test engineer who finds himself involved in type testing or elaborate acceptance tests is a study of this Standard (in either version). It is in six parts, briefly summarized below.

Part 1. Standard reference conditions, declarations of power, fuel and lubricating oil consumptions and test methods.

In this Part standard atmospheric conditions are specified as follows:

atmospheric pressure 1 bar (= 750 mm Hg)
temperature 25°C (298 K)
relative humidity 30%

the American SAE standards specify the same conditions (for marine engines standard conditions are specified as 45°C and 60% relative humidity)

Specific fuel consumptions should be related to a Lower Calorific Value of 42 700 kJ/kg: alternatively the actual LCV should be quoted.

The various procedures for correcting power and fuel consumption to these conditions are laid down and a number of examples are given. These procedures are extremely complicated and not easy to use: they involve 45 different symbols, 17 equations and 10 look-up tables. The procedures in the SAE standards listed above are only marginally less complex. Rigorous application of these rules involves a good deal of work and for everyday test work the 'correction factors' applied to measurements of combustion air consumption (the prime determinant of maximum engine power output) described in Chapter 10, p.183, will be found quite adequate.

Definitions are given for a number of different kinds of rated power: continuous, overload, service, ISO, etc. and a long list of auxiliaries which may or may not be driven by the engine is provided. It is specified that in any declaration of brake power the auxiliaries operating during the test should be listed and, in some cases, the power absorbed by the auxiliary should be given.

The section on test methods gives much detailed advice and instruction, including tables listing measurements to be made, functional checks and tests for various special purposes. Finally, Part 1 gives a useful check list of information to be supplied by the customer and by the manufacturer, including the contents of the test report.

Part 3. Test Measurements

This Part discusses 'accuracy' in general terms but consists largely in a tabulation of all kinds of measurement associated with engine testing with, for each, a statement of the 'Permissible deviation'. The definition of this term is extremely limited: it defines the range of values over which successive measurements made during a particular test are allowed to vary for the test to be valid. These limits are by no means tight: e.g. $\pm 3\%$ for power, $\pm 3\%$ for specific fuel consumption. They thus have little relevance to the general subject of accuracy.

Part 4. Speed Governing

This Part, which has been considerably elaborated in the 1997 Issue, specifies four levels of governing accuracy, M1 to M4 and gives detailed instructions for carrying out the various tests.

Part 5. Torsional vibrations

This Part is mainly concerned with defining the division of responsibility between the engine manufacturer, the customer and the supplier of the 'set', or machinery to be driven by the engine, e.g. a generator, compressor or ship propulsion system. In general the supplier of the set is regarded as responsible for calculations and tests.

Part 6. Specification of overspeed protection

This Part defines the various parameters associated with an overspeed protection system. The requirements are to be agreed between engine manufacturer and customer and it is recommended that reset after overspeed should be manual.

Part 7. Codes for Engine Power

This Part defines various letter codes, e.g. ICN for ISO Standard Power in English, French, Russian and German.
    It will be clear that this Standard gives much valuable guidance regarding many aspects of engine testing.

Summary

This chapter should be read in conjunction with Chapter 20, since the underlying theme is that of accuracy in the measurement of the key factors in engine performance: torque, power, fuel and lubricant consumption. The relationship between the various measurements is discussed, the appropriate instrumentation described and the process of dynamometer calibration explained in detail. The different kinds of acceptance and type tests are described and the standard British procedure for running such tests is summarized.

References

1. Kaye, G.W.C. and Laby, T.H. (1973) *Tables of physical and chemical constants*, Longmans, London.
2. Box, G.E.P. *et al.* (1978) *Statistics for experimenters*, Wiley, New York.
3. Neale, M.J. (1995) *The Tribology Handbook*, Butterworth, London.
4. Johren, P-W. and Newman, B.A. (1988) *Evaluating the oil consumption behaviour of reciprocating engines in transient operation*, SAE Paper 880098.
5. BS 5514 Parts 1 to 6 *Reciprocating Internal Combustion Engines: Performance*.
6. ISO 3046 *Reciprocating Internal Combustion Engines: Performance*
7. BS AU 141a *Specification for the Performance of Diesel Engines for Road Vehicles*.
8. Rulebook, Chapter 8, Lloyd's Register of Shipping, London.
9. SAE Standard *Engine power test code – spark ignition and compression ignition – gross power rating* SAE J1995 Jun 90.
10. SAE Standard *Engine power test code – spark ignition and compression ignition – net power rating* SAE J1349 Jun 90.

10 Measurement of combustion air consumption

The accurate measurement of the air consumption, commonly referred to as the combustion air, of an internal combustion engine is a matter of considerable difficulty, but of great importance. The influence of the condition of the air entering the engine on various aspects of engine performance is discussed and 'correction factors' defining these effects quantitatively are derived. The theory of various methods of measurement is given and the limitations of each method described.

Properties of air

Air is a mixture of gases with the following approximate composition:

	By mass	By volume
Oxygen, O_2	23.15%	20.95%
Nitrogen, N_2 + rare gases, mostly argon	76.85%	79.05%
	100%	100%

plus water vapour, variable, usual range 0.2% to 2.0% of volume of dry air.

The amount of water vapour present depends on temperature and prevailing atmospheric conditions. It can have an important influence on engine performance, notably on exhaust emissions, but for all but the most precise work its influence on air flow measurement may be neglected.

The relation between the pressure, specific volume and density of air is described by the gas equation

$$p_a \times 10^5 = \rho R T_a \tag{1}$$

where R, the gas constant, has the value for air

$$R = 287 \text{ J/kg K}$$

A typical value for air density in temperate conditions at sea level would be:

$$\rho = 1.2 \text{ kg/m}^3$$

Air consumption and engine performance

The internal combustion engine is essentially an 'air engine' in that air is the working fluid; the function of the fuel is merely to supply heat. There is seldom any particular technical difficulty in the introduction of sufficient fuel into the working cylinder but the attainable power output is strictly limited by the charge of air that can be aspirated.

It follows that the achievement of the highest possible volumetric efficiency is an important goal in the development of high-performance engines, and the design of inlet and exhaust systems, valves and cylinder passages represents a major part of the development programme for engines of this kind.

It is a surprising fact that, in the authors' experience, the effect of the condition of the air drawn in by the engine from the surrounding atmosphere on performance is often ignored, at least in routine tests. As already pointed out, p.180. The standard methods of taking into account the effects of charge air condition as laid down in European and American Standards are complex and difficult to apply and are mostly used to correct the power output of engines undergoing acceptance or type tests. A simplified treatment, adequate for most routine test purposes, is given below.

The condition of the air entering the engine is a function of the following parameters:

pressure
temperature
moisture content
impurities

For the first three factors standard conditions, according to European and American practice are:

atmospheric pressure 1 bar (= 750 mm Hg)
temperature 25°C (298 K)
relative humidity 30%

Atmospheric pressure

Since the volumetric efficiency of an engine tends to be largely independent of the air supply pressure the mass of air consumed tends to vary directly with the density, which is itself proportional to the absolute pressure, other conditions remaining unchanged. Since the standard atmosphere = 1 bar we may write:

$$\rho_t = \rho_n p \tag{2}$$

where: ρ_t = density under test conditions, kg/m^3
ρ_n = density under standard conditions, kg/m^3
p = atmospheric pressure under test conditions, bar

It follows that a change of 1% or 7.5 mmHg corresponds to a change in the mass of air entering the engine of 1%. For most days of the year the (sea level) atmospheric pressure will lie within the limits 750 mmHg ± 3%, say between 775 mmHg and 730 mmHg, with a corresponding percentage variation in charge air mass of 6%.

It is common practice to design the test cell ventilation system to maintain a small negative pressure in the cell, to prevent fumes entering the control room, but the level of depression is unlikely to exceed about 25 mm water gauge, equivalent to a change in barometer reading of less than 2 mmHg. Clearly the effect on combustion air flow, if the engine is drawing air from within the cell, is negligible.

The barometer also falls by about 86 mmHg for an increase in altitude of 1000 m (the rate decreasing with altitude), see Table 10.1. This indicates that the mass of combustion air falls by about 1% for each 90 m (300 ft) increase in altitude, a very significant effect.

Variations in charge air pressure have an important 'knock on' effect: the pressure in the cylinder at the start of compression will in general vary with the air supply pressure and the pressure at the end of compression will change in the same proportion. This can have a significant effect on the combustion process.

Temperature

Variations in the temperature of the air supply have an effect of the same order of magnitude as variations in barometric pressure within the range to be expected in test cell operation. Air density varies inversely with its absolute temperature:

$$\rho_t = \rho_n \cdot \frac{298}{(t_t + 273)} \tag{3}$$

Table 10.1. *Variation in atmospheric pressure with height above sea level*

Altitude m	Fall in pressure bar
0	0
500	0.059
1000	0.115
1500	0.168
2000	0.218
3000	0.312
4000	0.397

where t_t = test temperature

However, the temperature at the start of compression determines that at the end. In the case of a naturally aspirated diesel engine with a compression ratio of 16:1 and an air supply at 25°C the charge temperature at the start of compression would typically be about 50°C. At the end of compression the temperature would be in the region of 530°C, increasing to about 560°C for an air supply temperature 10°C higher. The level of this temperature can have a significant effect on such factors as the NO_x content of the exhaust, which is very sensitive to peak combustion temperature. The same effect applies, with generally higher temperatures, to turbo-charged engines.

The effect of atmospheric pressure, altitude and charge air temperature on the mass of the air charge are summarized in Fig. 10.1.

Compared with the effects of pressure and temperature the influence of the relative humidity of the air supply on the air charge is relatively small, except at high air temperatures. The moisture content of the combustion air does however exert a number of influences on performance. Some of the thermo-dynamic properties of moist air have been discussed under the heading of psychrometry, p.62, but certain other aspects of the subject must now be considered. The important point to note is that unit volume of moist air contains less oxygen in a form available for combustion than the same volume of dry air under the same conditions of temperature and pressure. Moist air is a mixture of air and steam: while the latter contains oxygen it is in chemical combination with hydrogen and is thus not available for combustion.

The European and American Standards specify a relative humidity of 30% at 25°C. This corresponds to a vapour pressure 0.01 bar, thus implying a dry air pressure of 0.99 bar and a corresponding reduction in oxygen content of 1% when compared with dry air at the same pressure. Fig 10.2 shows the variation in dry air volume expressed as a percentage adjustment to the standard condition with temperature for relative humidities ranging from 0% (dry) to 100% (saturated).

The effect of humidity becomes much more pronounced at higher temperatures: thus at a temperature of 40°C the charge mass of saturated air is 7.4% less than for dry air. The effect of temperature on moisture content (humidity) should be noted particularly, since the usual method of indicating specified moisture content for a given test method, by specifying a value of relative humidity, can be misleading. Figure 10.2 indicates that at a temperature of 0°C the difference between 0% (dry) and 100% (saturated) relative humidity represents a change of only 0.6 % in charge mass; at −10°C it is less than 0.3%. It follows that it is barely worthwhile adjusting the humidity of combustion air at temperatures below (perhaps) 5°C.

The moisture content of the combustion air has a significant effect on the formation of NO_x in the exhaust of diesel engines. The SAE procedure for measurement of diesel exhaust emissions[6] gives a rather complicated

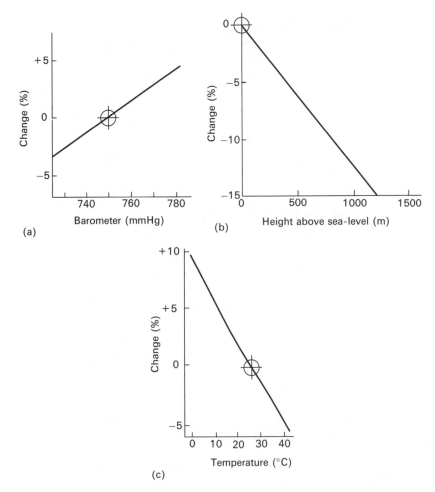

Figure 10.1.   *Effect of various factors on mass of charge air (a) effect of baro-metric pressure (b) effect of altitude (c) effect of charge air temperature. Standard conditions: 750 mm Hg, 25°C, 30% relative humidity*

expression for correcting for this. This indicates that, should the NO_x measurement be made with completely dry air, a correction of the order of $+15\%$ should be made to give the corresponding value for a test with moisture content of 60% relative humidity.

Finally, it should be mentioned that pollution of the combustion air can result in a reduction in the oxygen available for combustion. A likely source of such pollution may be the ingestion of exhaust fumes from other engines or from neighbouring industrial processes such as paint plant and care should also be taken in the siting of air intakes and exhaust discharges to ensure that under

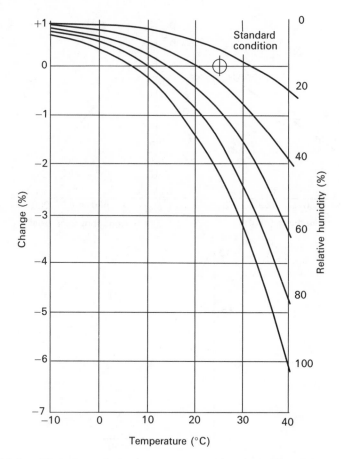

Figure 10.2.   *Variation of air charge mass with relative humidity at different temperatures*

unfavourable conditions there cannot be unintended exhaust recirculation. Exhaust gas has much the same density as air and a free oxygen content that can be low or even zero, so that, approximately, 1% of exhaust gas in the combustion air will reduce the available oxygen in almost the same proportion.

These various adjustments or 'corrections' to the charge mass are of course cumulative in their effect and the following example illustrates the magnitudes involved. Consider:

A a hot, humid summer day with the chance of thunder
B a cold, dry winter day of settled weather

	A	B
pressure	0.987 bar, 740 mmHg	1.027 bar, 770 mmHg
temperature	35°C	10°C
relative humidity	80%	40%

pressure, eq. (10.2)

$$\rho_n = \rho_t x \qquad \frac{1}{0.987} = 1.0135 \qquad \frac{1}{1.027} = 0.9740$$

temperature, eq. (10.3)

$$\rho_n = \rho_t x \qquad \frac{308}{298} = 1.0336 \qquad \frac{283}{298} = 0.9497$$

relative humidity from Fig. 10.2

$$+ 3.48\% = 1.0348 \qquad\qquad 0\%$$

total adjustment   1.0840,   + 8.4%   0.9233,   −7.7%

This 'adjustment' indicates the factor by which the observed power should be multiplied to indicate the power to be expected under 'standard' conditions. We see that under hot, humid, stormy conditions the power may be reduced by as much as 15% compared with the power under cold anti-cyclonic conditions, a by no means negligible adjustment.

It should be pointed out that the power adjustment calculated above is based on the assumption that charge air mass *directly* determines the power output but this would only be the case if the air/fuel ratio were rigidly controlled (as in spark ignition engines with precise stoichiometric control). In most cases a reduction, for example, in charge air mass due to an increase in altitude will result in a reduction in the air/fuel ratio, perhaps with an increase in exhaust smoke, but not necessarily in a reduction in power. The correction factors laid down in the various Standards, p.182, take into account the differing responses of the various types of engine to changes in charge oxygen content.

It should also perhaps be pointed out that the effects of water injection are quite different from the effects of humidity already present in the air. Humid air is a mixture of air and steam: the latent heat required to produce the steam has already been supplied. When, as is usually the case, water is injected into air leaving the turbo-charger at a comparatively high temperature, the cooling effect associated with the evaporation of the water achieves an increase in charge density which much outweighs the decrease associated with the resulting steam.

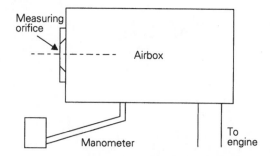

Figure 10.3.   *Airbox with measuring orifice and manometer*

The airbox method of measuring air consumption

The simpler methods of measuring air consumption involve drawing the air through some form of measuring orifice, Fig. 10.3, and measuring the pressure drop across the orifice. It is good practice to limit this drop to not more than about 125 mm H_2O (1200 Pa). For pressures less than this air may be treated as an incompressible fluid, with much simplification of the air flow calculation.

The velocity U developed by a gas expanding freely under the influence of a pressure difference Δp, if this difference is limited as above, is given by:

$$\frac{\rho U^2}{2} = \Delta p; U = \sqrt{\frac{2\Delta p}{\rho}} \tag{4}$$

Typically, air flow is measured by means of a sharp-edged orifice mounted in the side of an airbox, coupled to the engine inlet and of sufficient capacity to damp out the inevitable pulsations in the flow into the engine, which are at their most severe in the case of a single cylinder 4-stroke engine, Fig. 10.4. In the case of turbocharged engines the inlet air flow is comparatively smooth and a well-shaped nozzle without an airbox will give satisfactory results.

The air flow through a sharp-edged orifice takes the form sketched in Fig. 10.5. The coefficient of discharge of the orifice is the ratio of the transverse area of the flow at plane a (the vena contracta) to the plan area of the orifice. Tabulated values of C_d are available[1], but for many purposes a value $C_d = 0.60$ may be assumed.

We may easily derive the volumetric flow rate of air through a sharp-edged orifice as follows:

Flow rate = coefficient of discharge × cross-sectional area of orifice × velocity of flow

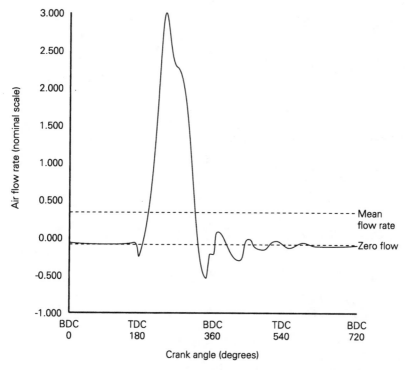

Figure 10.4. *Induction air flow, single cylinder four-stroke diesel engine*

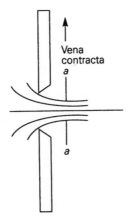

Figure 10.5. *Flow through a sharp edged orifice*

$$Q = C_d \frac{\pi d^2}{4} \sqrt{\frac{2\Delta p}{\rho}} \qquad (5)$$

Noting from eq. (3), that

$$\rho = \frac{p_a \times 10^5}{RT_a}$$

and assuming that Δp is equivalent to h mm H_2O we may write

$$Q = C_d \frac{\pi d^2}{4} \sqrt{\frac{2 \times 9.81h \times 287T}{p_a \times 10^5}} \qquad (6)$$

$$Q = 0.1864 C_d d^2 \sqrt{\frac{hT_a}{p_a}} \, m^3/s \qquad (6a)$$

To calculate the mass rate of flow, note that:

$$\dot{m} = \rho Q = \frac{p_a Q}{RT_a}$$

giving, from eq. (6):

$$\dot{m} = C_d \frac{\pi d^2}{4} \sqrt{\frac{2 \times 9.81 h p_a \times 10^5}{287 T_a}} \qquad (7)$$

$$\dot{m} = 64.94 C_d d^2 \sqrt{\frac{h p_a}{T_a}} \qquad (7a)$$

Equations (6a) and (7a) give the fundamental relationship for measuring air flow by an orifice, nozzle, or venturi.

Sample calculation

If

$$C_d = 0.6$$
$$d = 0.050 \text{ m}$$
$$h = 100 \text{ mm } H_2O$$
$$T_a = 293 \text{ K}(20°C)$$
$$p_a = 1.00 \text{ bar}$$

then

Table 10.2.   *Approximate flow rates for orifices*

Orifice dia. mm	Q (m^3/s)	$\dot{m}$ (kg/s)
10	0.002	0.002
20	0.008	0.009
50	0.048	0.057
100	0.19	0.23
150	0.43	0.51

$$Q = 0.04786 \text{ m}^3/\text{s} \ (1.69 \text{ ft}^3/\text{s})$$
$$\dot{m} = 0.05691 \text{ kg/s} \ (0.1255 \text{ lb/s})$$

To assist in the selection of orifice sizes, Table l0.2 gives approximate flow rates for orifices under the following standard conditions:

$$h = 100 \text{ mm H}_2\text{O}$$
$$T_a = 293 \text{ K } (20°\text{C})$$
$$p_a = 1.00 \text{ bar}$$

A disadvantage of flow measurement devices of this type is that the pressure difference across the device varies with the square of the flow rate. It follows that a turn down in flow rate of 10 : 1 corresponds to a reduction in pressure difference of 100 : 1, implying insufficient precision at low flow rates. It is good practice, when a wide range of flow rates is to be measured, to select a range of orifice sizes, each covering a turn down of not more than 2.5 : 1 in flow rate.

Air consumption of engines : approximate calculation

The air consumption of an engine may be calculated from:

$$V = \eta_v \frac{V_s}{K} \frac{n}{60} \text{m}^3/\text{s} \tag{8}$$

where $K = 1$ for a two-stroke and 2 for a four-stroke engine.

For initial sizing of the measuring orifice η_v, the ratio of the volume of air aspirated per stroke to the volume of the cylinder, may be assumed to be about 0.8 for a naturally aspirated engine and up to about 2.5 for supercharged engines.

Sample calculation

Single-cylinder 4-stroke engine, swept volume 0.8 litre, running at a maximum of 3000 rev/min, naturally aspirated:

$$V = 0.8\frac{0.0008}{2}\frac{3000}{60} = 0.16 \text{ m}^3/\text{s}$$

Suitable orifice size, from Table 10.1, 30 mm.

Minimum size of airbox

It is necessary to ensure that the airbox has a certain minimum volume, which is a function of engine size and other characteristics, if flow through the orifice is to be sufficiently smooth for this method of measurement to be reliable. In general, the larger the engine, the slower the speed and the smaller the number of cylinders the larger is the necessary size of the airbox.

An analysis of the problem was made by Kastner[2], who derived the following expression for the minimum desirable size of airbox:

$$V_b = \frac{417 \times 10^6 \times K^2 d^4}{N_c V_s n_{min}^2} \tag{9}$$

In this expression n_{min} is the minimum engine speed at which accurate measurements are required.

Sample calculation

The engine that is the subject of the previous calculation running at 1000 rev/min.

$$V_b = \frac{417 \times 10^6 \times 2^2 \times 0.030^4}{1 \times 0.0008 \times 1000^2} = 1.70 \text{ m}^3 \text{ (60 ft}^3\text{)}$$

Connection of airbox to engine inlet

It is essential that the configuration of the connection between the airbox and the engine inlet should model as closely as possible the configuration of the air intake arrangements in service. This is because pressure pulsations in the inlet can have a powerful influence on engine performance, in terms of both volumetric efficiency and pumping losses.

The resonant frequency of a pipe of length l, open at one end and closed at the other $= a/4l$, where a is the speed of sound, roughly 330 m/s. Thus an inlet connection 1 m long would have a resonant frequency of about 80 Hz. This corresponds to the frequency of intake valve opening in a 4 cylinder 4-stroke engine running at 2400 rev/mm. Clearly such an intake connection could disturb engine performance at this speed. In general the intake connection should be as short as possible.

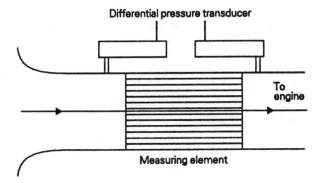

Figure 10.6. *Viscous flow airmeter*

The viscous flow airmeter

The viscous flow airmeter, invented by Alcock and Ricardo in 1936, is the most widely used alternative to the airbox and orifice method of measuring air flow. In this device, Fig. 10.6, the measuring orifice is replaced by an element consisting of a large number of small passages, generally of triangular form. The flow through these passages is substantially laminar, with the consequence that the pressure difference across the element is approximately proportional to the velocity of flow, rather than to its square, as is the case with a measuring orifice.

This has two advantages. First average flow is proportional to average pressure difference, implying that a measurement of average pressure permits a direct calculation of flow rate, without the necessity for smoothing arrangements. Secondly the acceptable turn-down ratio is much greater.

The traditional version of the viscous flow airmeter employs various methods, not entirely successful, to measure the average of the pressure difference. An analysis by Stone[3] has shown that pressure difference is not exactly proportional to flow rate, and has indicated a precise method of obtaining a true average value of pressure difference. These changes, now commercially available, have greatly improved the accuracy of the viscous flow airmeter, and have rendered possible its use in transient conditions. The flowmeter must be calibrated against a standard device, such as a measuring orifice.

Other methods of measuring air consumption – the Lucas–Dawe air mass flowmeter

This device, Fig. 10.7, depends for its operation on the corona discharge from an electrode coincident with the axis of the duct through which the air is

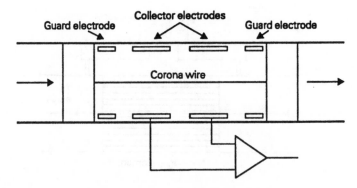

Figure 10.7.   *Lucas–Dawe air mass flowmeter*

flowing. Air flow deflects the passage of the ion current to two annular electrodes and gives rise to an imbalance in the current flow that is proportional to air flow rate.

An advantage of the Lucas–Dawe flowmeter is its rapid response to changes in flow rate, of the order of 1 ms, making it well suited to transient flow measurements, but it is sensitive to air temperature and humidity, and requires calibration against a standard.

Positive displacement flowmeters

As rotation occurs successive pockets of air are transferred from the suction to the delivery side of the flowmeter, and flow rate is proportional to rotor speed. Some of these flowmeters operate on the principle of the Roots blower. An ingenious version, in which a rotor and vanes interact, is shown in Fig. 10.8.

Advantages of the positive displacement flowmeter are accuracy, simplicity and good turn-down ratio. Disadvantages are cost, bulk, relatively large pressure drop, and sensitivity to contamination in the flow.

Measurement of crankcase blow-by

Heywood[4] gives a detailed discussion of the effect of 'crevice flow' on engine performance. By this is meant the flow of gas into and out of the clearances between piston top land, ring grooves and cylinder bore. In a typical automotive gasolene engine these can amount to 3% of the combustion chamber volume. Since in a spark-ignition engine flow into these devices consists of unburned mixture which emerges during the expansion stroke too late to be burned this is a major source of HC emissions and also represents a loss of power.

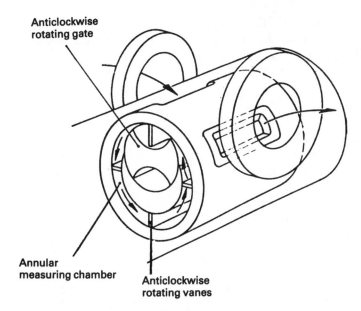

Anticlockwise
rotating gate

Annular
measuring chamber

Anticlockwise
rotating vanes

Figure 10.8.   *Positive displacement flowmeter*

Some of this gas will leak past the rings and piston skirt in the form of blow-by into the crankcase. It is then vented back into the induction manifold and to this extent reduces the HC emissions and fuel loss, but has a very adverse effect on the lubricant. Blow-by volume when the engine is running correctly is largely governed by the size of ring gaps, and a typical rate for a passenger car engine would be in the range $0.007-0.025$ m^3/min ($0.25-0.90$ ft^{3}min).

Crankcase blow-by is a significant indicator of engine condition and should preferably be monitored during any extended test sequence. An increase in blow-by can be a symptom of various problems such as incipient ring sticking, bore polishing or deficient cylinder bore lubrication.

Blow-by is measured by fitting a small capacity flowmeter to the crankcase vent. Any flowmeter intended for gas flows of a few litres per minute should be suitable and good results have been obtained using a simple domestic gas meter.

Measurement of air consumption: summary of procedure

(1) Calculate air consumption of engine over a representative range of conditions from eq. (8).
(2) Select a suitable range of orifice sizes from Table 10.2 or eq. (6a).
(3) Select minimum size of airbox from eq. (9).

(4) Design a suitable inlet connection.
(5) Alternatively employ a viscous flow or positive displacement airmeter directly connected to engine inlet.

Notation

atmospheric pressure	p_a bar
under test conditions	p bar
atmospheric temperature	T_a K
test temperature	$t_t\,°C$
density of air	ρ kg/m^3
under standard conditions	ρ_n kg/m^3
under test conditions	ρ_t kg/m^3
pressure difference across orifice	Δp Pa, h mmH$_2$O
velocity of air at contraction	U m/s
coefficient of discharge of orifice	C_d
diameter of orifice	d m
volumetric rate of flow of air	Q m^3/s
mass rate of flow of air	$\dot{m}$ kg/s
constant, 1 for 2-stroke and 2 for 4-stroke engines	K
engine speed	n rev/min
number of cylinders	N_c
swept volume, total	V_s m^3
air consumption rate	V m^3/s
volumetric efficiency of engine	η_v
volume of airbox	V_b m^3
gas constant	R J/kg K

References

1. BS 1042 *Measurement of Fluid Flow in Closed Conduits: Section 1.1, Specification for Square-edged Orifice Plates, Nozzles and Venturi Tubes inserted in Circular cross-section; Conduits running full; Section 1.4, Guide to the use of Devices specified in Sections 1.1 and 1.2.*

2. Kastner, L.J. (1947) The airbox method of measuring air consumption, *Proc. I. Mech. E.,* **157**.

3. Stone, C.R. (1989) *Airflow Measurement in Internal Combustion Engines,* SAE Technical Paper, Series 890242.

4. Heywood, J.B. (1988) *Internal Combustion Engine Fundamentals,* McGraw-Hill, Maidenhead.

5. BS 5514 Parts 1 to 6 *Reciprocating Internal Combustion Engines: Performance*
6. *Measurement of carbon dioxide, carbon monoxide, and oxides of nitrogen in diesel exhaust* SAE J177 Apr 82.

Further reading

Plint, M.A. and Böswirth, L. (1978) *Fluid Mechanics: A Laboratory Course*, Griffin, London.
BS 7405 *Guide to Selection and Application of Flowmeters for the Measurement of Fluid Flow in Closed Conduits.*

11 Thermal efficiency and measurement of heat losses – ideal standard cycles

Effect of compression ratio on performance

This chapter deals with the measurements and calculations necessary to determine the energy balance and thermal performance of an internal combustion engine. To provide a framework for an interpretation of these observations a brief account is given of the basic theory.

One ultimate measure of the performance of an internal combustion engine is the proportion of the heat of combustion of the fuel that is turned into useful work at the engine coupling. The thermal efficiency at full load of internal combustion engines ranges from about 20% for small gasoline engines up to more than 50% for large slow-running diesel engines. It is worth pointing out that the thermal efficiency of very large steam power stations is unlikely to exceed 35%; the slow-running marine diesel engine is the most efficient means available of turning the heat of combustion of fuel into mechanical power.

It is useful to have some idea of the theoretical maximum thermal efficiency that is possible, as this sets a target for the engine developer. Fortunately theoretical thermodynamics allows us, within certain limitations, to predict this maximum value.

The proportion of the heat of combustion that is not converted into useful work appears elsewhere: in the exhaust gases, in the cooling medium and as convection and radiation from the hot surfaces of the engine.

There may in addition be appreciable losses in the form of unburned or late burning fuel. It is important to be able to evaluate these various losses. Of particular interest are losses from the hot gas in the cylinder to the containing surfaces, since these directly affect the indicated power of the engine. The so-called 'adiabatic engine' seeks to minimize these particular losses.

Some of the heat carried away in the exhaust gas may be converted into useful work in a turbine or used for such purposes as steam generation or the production of hot water.

Fundamentals

Calorific value of fuels

The calorific value of a fuel is defined in terms of the amount of heat liberated when a fuel is burned completely in a calorimeter. Detailed methods and definitions are given in ref. 1, but for the present purpose the following is sufficient.

Since all hydrocarbon fuels produce water as a product of combustion part of these products (the exhaust gas in the case of an i.c. engine) consists of steam. If, as is the case in a calorimeter, the products of combustion are cooled to ambient temperature, this steam condenses, and in doing so gives up its latent heat.

The corresponding measure of heat liberated is known as the higher or gross calorific value (also known as gross specific energy). If no account is taken of this latent heat we have the lower or net calorific value (also known as net specific energy). Since there is no possibility of an internal combustion engine making use of the latent heat, it is the invariable practice to define performance in terms of the lower calorific value C_1.

Table 11.1 shows values of the lower calorific value and density for some typical fuels.

Gaseous fuels

These fuels, which have favourable emissions characteristics, are becoming of increasing importance[2,3].

Table 11.1. *Properties of liquid fuels*

	Lower calorific value* (MJ/kg)	Stoichiometric air/fuel ratio	Density (kg/l)
Gasolene	43.8	14.6	0.74
Gas oil	42.5	14.8	0.84
Methanol	19.9	6.46	0.729
Ethanol	27.2	8.94	0.79
Light fuel oil	40.6		0.925
Medium fuel oil	39.9	14.4	0.95
Heavy fuel oil	39.7		0.965

*BS 5514 Part 1 *Reciprocating Internal Combustion Engines*: Performance specifies a standard Lower Calorific Value for distillate fuels as 42.7 MJ/kg

Natural gas (NG), also sometimes described as compressed natural gas (CNG) and, when transported in bulk at very low temperature, as liquefied natural gas (LNG)

Natural gas deposits exist either as free gas or in association with crude oil.

Natural gas consists mainly of methane but, having evolved from organic deposits, invariably contains some higher hydrocarbons and traces of N_2 and CO_2. Composition varies considerably from field to field and the reader involved in work on natural gas should establish the particulars of the gas with which he is concerned. North Sea Gas has become the standard gaseous fuel in the UK and its approximate properties are shown in Table 11.2.

Liquefied petroleum gas (LPG or LP-gas)

LPG is a product of the distillation process of crude oil or a condensate from wet natural gas. It consists largely of propane and, unlike natural gas, can be stored in liquid form at moderate pressures. In view of its good environmental properties (low unburned hydrocarbon emissions, low CO, virtually no particulate emissions and no sulphur) it is finding favour as a vehicle engine fuel. NO_x emissions tend to be higher than for gasolene, owing to its high combustion temperature. It has a high octane number, RON 110, which permits higher compression ratios. Approximate properties are shown in Table 11.2.

Ideal standard cycles: effect of compression ratio

Many theoretical cycles for the internal combustion engine have been proposed, some of them taking into account such factors as the exact course of the combustion process, the variation of the specific heat of air with temperature and the effects of dissociation of the products of combustion at high temperature. However, all these cycles merely modify the predictions of the cycle we shall be considering, generally in the direction of reduced attainable efficiency, without much changing the general picture. For a detailed discussion see Heywood[4].

The air standard cycle, also known as the Otto cycle, is shown in Fig. 11.1. It consists of four processes, forming a complete cycle, and is based on the following assumptions:

(1) The working fluid throughout the cycle is air, and this is treated as a perfect gas.
(2) The compression process 1–2 and the expansion process 3–4 are both treated as frictionless and adiabatic (without heat loss).

Table 11.2. *Approximate properties of typical gaseous fuels[3]*

Natural gas (North Sea gas)	
Methane CH_4	93.3%
Higher hydrocarbons	4.6%
N_2, CO_2	2.1%
Lower calorific value*	48.0 MJ/kg
Stoichiometric air/fuel ratio	14.5:1
Approximate density (gas at 0°C)	$0.79 kg/m^3$
Liquefied petroleum gas (LPG)	
Propane, C_3H_8	90%
Butane C_4H_{10}	5%
Unsaturates	5%
Lower calorific value	46.3 MJ/kg
Stoichiometric air/fuel ratio	15.7:1
Approximate density (gas at 0°C)	$2.0 kg/m^3$

* It should be noted that some gas suppliers commonly quote higher as opposed to lower calorific value. In the case of methane, with its high H/C ratio, the difference is nearly 10%.

(3) In place of heat addition by internal combustion phase 2–3 of the process is represented by the addition of heat from an external source, the volume remaining constant.
(4) The exhaust process is replaced by cooling at constant volume to the initial temperature.

It will be evident that these conditions differ considerably from those

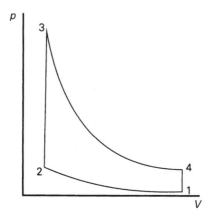

Figure 11.1. *Air standard cycle*

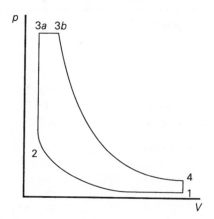

Figure 11.2. *Limited pressure cycle*

encountered in an engine; nevertheless the thermodynamic analysis, which is very simple, gives useful indications regarding the performance to be expected from an internal combustion engine, in particular with regard to the influence of compression ratio.

The air standard cycle efficiency is given by:

$$\eta_{as} = 1 - \frac{1}{R^{\gamma-1}} \qquad (1)$$

The course of events in an engine cylinder departs from this theoretical pattern in the following main respects:

(1) Heat is lost to the cylinder walls, reducing the work necessary to compress the air, the rise in temperature and pressure during combustion, and the work performed during expansion.

(2) Combustion, particularly in the diesel engine, does not take place at constant volume, resulting in a rounding of the top of the diagram, point 3, and a reduction in power. A better standard of reference for the diesel engine is the limited pressure cycle, Fig. 11.2, for which the efficiency is given by the expression:

$$\eta = 1 - \frac{1}{R^{\gamma}-1} \left[\frac{\alpha\beta^{\gamma}-1}{\alpha^{\gamma}(\beta-1)+\alpha-1} \right] \qquad (1a)$$

where

$$\alpha = \frac{p_3}{p_{2_a}}, \qquad \beta = \frac{V_{3_b}}{V_{3_a}}$$

this reduces to eq. (1) when $\beta = 1$.

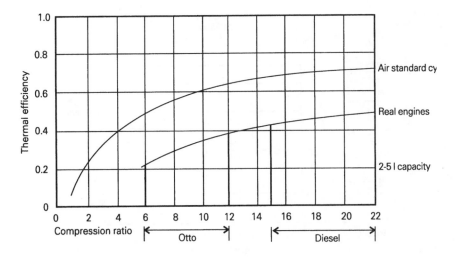

Figure 11.3.   *Variation of air standard cycle efficiency with compression ratio*

(3)   The properties of air, and of the products of combustion, do not correspond to those of an ideal gas, resulting in a smaller power output than predicted.

(4)   The gas exchange process is ignored in the standard cycle.

Figure 11.3 shows the variation of air standard cycle efficiency with compression ratio, and shows the range of this ratio for spark ignition and diesel engines. It is clearly desirable to use as high a compression ratio as possible. Also shown is the approximate indicated thermal efficiency to be expected from gasoline and diesel engines of 2–5 litre swept volume. Larger engines, and in particular large slow-speed diesel engines, can achieve significantly higher efficiencies, mainly because heat losses from the cylinder contents become less in proportion as the size of the individual cylinders increases.

The energy balance of an internal combustion engine

The distribution of energy in an internal combustion engine is best considered in terms of the steady flow energy equation, combined with the concept of the control volume. In Fig. 11.4 an engine is shown, surrounded by the control surface. The various flows of energy into and out of the control volume are shown.

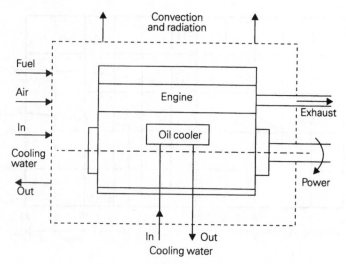

Figure 11.4.   *Control volume of i.c. engine showing energy flows*

In

- fuel, with its associated heat of combustion
- air, consumed by the engine

Out

- power developed by the engine
- exhaust gas
- heat to cooling water or air
- convection and radiation to the surroundings

The *steady flow energy equation* gives the relationship between these quantities, and is usually expressed in kilowatts:

$$H_1 = P_s + (H_2 - H_3) + Q_1 + Q_2 \tag{2}$$

in which the various terms have the following meanings:

H_1 = combustion energy of fuel = $\dot{m}_f C_L \times 10^3$
P_s = power output of engine
H_2 = enthalpy of exhaust gas* = $(\dot{m}_f + \dot{m}_a)C_p T_e$
H_3 = enthalpy of inlet air = $\dot{m}_a C_p T_a$
Q_1 = heat to cooling water† = $\dot{m}_w C_w (T_{2w} - T_{1w})$
Q_2 = convection and radiation.

---

* For a definition of enthalpy see, for example, refs 4, 5.
† This may also include heat transferred to the lubricating oil and subsequently transferred to the cooling water in a separate oil cooler.

This assumes that the specific heat of the exhaust gas, the mass of which is the sum of the masses of air and fuel supplied to the engine, is equal to that of air. This is not strictly true, but permits an approximate calculation to be made if the temperature of the exhaust gas is measured (exact measurement of exhaust temperature is no simple matter, see Chapter 20).

Note that it is not possible to show the indicated power directly in this energy balance since the difference between it and the power output P_s, representing friction and other losses, appears elsewhere as part of the heat to the cooling water Q_1 and other losses Q_2.

Measurement of heat losses: heat to exhaust

If air and fuel flow rates, also exhaust temperature, are known, this may be calculated approximately, see H_2 above. For an accurate measurement of exhaust heat, use is made of an exhaust calorimeter, Fig. 11.5. This is an air-to-water heat exchanger in which the exhaust gas is cooled to a moderate temperature and the heat content measured from observation of cooling water flow rate and temperature rise.

The expression for H_2 becomes:

$$H_2 = \dot{m}_c C_w (T_{2c} - T_{1c}) + (\dot{m}_f + \dot{m}_a) C_p T_{co} \tag{3}$$

The rate of flow of cooling water through the calorimeter should be regulated so that the temperature of the gas leaving the calorimeter, T_{co}, does not fall below about 60°C (333 K). This is approximately the dew point temperature for exhaust gas: at lower temperatures the steam in the exhaust will start to condense, giving up its latent heat, see section on Calorific Value of Fuels.

Sample calculation: analysis of an engine test

Table 11.3 is an analysis, based on eq. (2), of one test point in a sequence of tests on a vehicle engine.

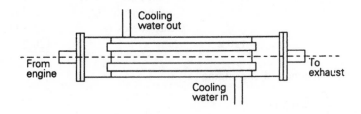

Figure 11.5. *Exhaust calorimeter*

Table 11.3. *Energy balance of a gasolene engine at full throttle (4 cylinder 4-stroke engine swept volume 1.7011)*

Engine speed	3125 rev/min
Power output	$P_s = 36.8$ kW
Fuel consumption rate	$\dot{m}_f = 0.00287$ kg/s
Air consumption rate	$\dot{m}_a = 0.04176$ kg/s
Lower calorific value of fuel	$C_L = 41.87 \times 10^6$ J/kg
Exhaust temperature	$T_c = 1066$ K (793°C)
Cooling water flow	$\dot{m}_w = 0.123$ kg/s
Cooling water inlet temperature*	$T_{1w} = 9.2$°C
Cooling water outlet temperature*	$T_{2w} = 72.8$°C
Inlet air temperature	$T_a = 292$ K (19°C)

* The engine was fitted with a heat exchanger. These are the temperatures of the primary cooling water flow to the exchanger.

Then noting that:

specific heat of air at constant pressure $C_p = 1.00$ kJ/kgK
specific heat of water $C_w = 4.18$ kJ/kgK

$$H_1 = 0.00287 \times 41.87 \times 10^3 \qquad = 120.2 \text{ kW}$$
$$P_s \qquad\qquad\qquad\qquad\qquad\qquad\qquad = 36.8 \text{ kW}$$

$$H_2 = (0.00287 + 0.04176) \times 1.00 \times 1066 \qquad = 47.6 \text{ kW}$$

$$H_3 = 0.04176 \times 1.00 \times 292 \qquad\qquad\qquad = 12.2 \text{ kW}$$

$$H_2 - H_3 \qquad\qquad\qquad\qquad\qquad\qquad\qquad = 35.4 \text{ kW}$$

$$Q_1 = 0.123 \times 4.18 \, (72.8 - 9.2) \qquad\qquad = 32.7 \text{ kW}$$

$$Q_2 \quad \text{(by difference)} \qquad\qquad\qquad\qquad = 15.3 \text{ kW}$$

We may now draw up an energy balance (quantities in kilowatts):

Heat of combustion H_1	120.2	Power output P_s	36.8 (30.6%)
		Exhaust $(H_2 - H_3)$	35.4 (29.5%)
		Cooling water Q_1	32.7 (27.2%)
		Other losses Q	15.3 (12.7%)
	120.2		120.2

The thermal efficiency of the engine

$$\eta_{th} = \frac{P_s}{H_1} = 0.306$$

The compression ratio of the engine $R = 8.5$, giving:

$$\eta_{as} = 1 - \frac{1}{8.5^{1.4-1}} = 0.575$$

The mechanical efficiency of this engine at full throttle was approximately 0.80, giving an indicated thermal efficiency of

$$\frac{0.306}{0.80} = 0.3825$$

This is approximately two thirds of the air standard efficiency.

Sample calculation: exhaust calorimeter

In the test analysed in Table 11.3 the heat content of the exhaust was also measured by an exhaust calorimeter with the following result:

cooling water flow	$= \dot{m}_c$	$= 0.139\,\text{kg/s}$
cooling water inlet temperature	$= T_{1c}$	$= 9.2°C$
cooling water outlet temperature	$= T_{2c}$	$= 63.4°C$
exhaust temperature leaving calorimeter	$= T_{co}$	$= 355\,\text{K}\ (82°C)$

Then from eq. (3):

$$H_2 = 0.139 \times 4.18\,(63.4 - 9.2) + (0.00287 + 0.04176) \times 1.00 \times 355$$

$$= 47.3 \text{ kW}$$

$H_3 = 12.2$ kW, as before, giving heat to exhaust $H_2 - H_3 = 35.1$ kW. This shows a satisfactory agreement with the approximate value derived from the exhaust temperature and air and fuel flow rates.

Energy balances: typical values

Figure 11.6 shows full power energy balances for several typical engines. Results for the gasolene engine analysed above are shown at (a). Table 11.4 shows energy balances in terms of heat losses per unit power output for various engine types. This will be found useful when designing such test cell services as cooling water and ventilation. They are expressed in terms of kW/kW power output.

These proportions depend on the thermal efficiency of the engine, and are only an approximate guide.

The role of indicated power in the energy balance

It should be observed that the indicated power output of the engine does not appear in any of our formulations of the energy balance. There is a good

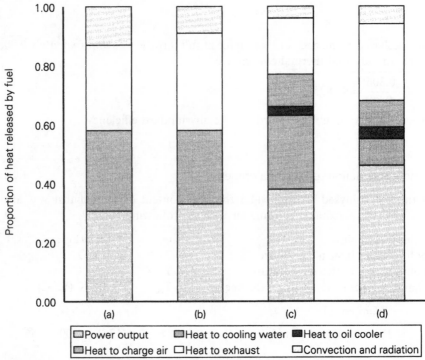

Figure 11.6. *Typical full power energy balances: (a) 1.71 gasolene engine; (b) 2.51 naturally aspirated diesel engine; (c) 200 kW medium speed turbocharged marine diesel; (d) 7.6 MW combined heat and power unit*

reason for this: the difference between indicated and brake power represents the friction losses in the engine and the power required to drive the auxiliaries, and it is impossible to allocate these between the various heat losses in the balance.

Table 11.4. *Energy balance (kW per kW power output)*

	Automotive gasolene	*Automotive diesel*	*Medium speed heavy diesel*
Power output	1.0	1.0	1.0
Heat to cooling water	0.9	0.7	0.4
Heat to oil cooler			0.05
Heat to exhaust	0.9	0.7	0.65
Convection and radiation	0.2	0.2	0.15
Total	3.0	2.6	2.2

Most of the friction losses between piston and cylinder will appear in the cooling water; bearing losses and the power required to drive the oil pump will appear mostly in the oil cooler while water pump losses will appear directly in the cooling water. An exact analysis is not possible.

Sample calculation: prediction of energy balance

Chapter 1 deals with the concept of the engine test cell as a thermodynamic system and the recommendation is made that at an early stage in designing a new test cell an estimate should be made of the various flows: fuel, air, water, electricity, heat and energy into and out of the cell.

In such cases it is usually necessary to make some assumptions as to the full power performance of the largest engine to be tested. This is possible on the basis of information given earlier and in this chapter and an example follows.

Prediction of energy balance: 250 kW turbo charged diesel engine at full power

Assume specific fuel consumption = 0.21 kg/kWh (LCV 40.6 MJ/kg; thermal efficiency 0.42)

Then following the general recommendations of Table 11.4 we may make an estimate, summarized in Table 11.5

The corresponding thermodynamic system is shown in Figure 11.7. This also shows rates of flow of fuel, air and cooling water, based on the following estimates:

Fuel flow

Assume fuel density 0.9 kg/litre

Table 11.5. *Energy balance, 250 kW turbocharged diesel engine*

In		Out	
Fuel	592 kW	Power	250 kW (42.2%)
		Heat to cooling water	110 kW (18.6%)
		Heat to oil cooler	15 kW (2.5%)
		Heat to exhaust	177 kW (29.9%)
		Convection and radiation	40 kW (6.8%)
	592 kW		592 kW

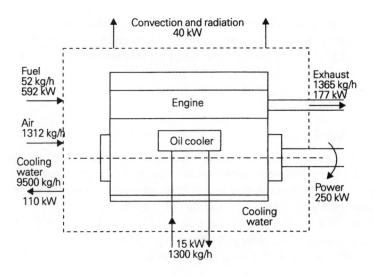

Figure 11.7.   *Control volume, 250 kW diesel engine, showing energy and fluid flows*

$$\text{fuel flow} = \frac{250 \times 0.21}{0.9} = 581/\text{h} \ (52.5 \ \text{kg/h})$$

Induction air flow

Assume full load air/fuel ratio = 25:1
Air flow = $250 \times 0.21 \times 25 = 1312.5$ kg/h
Taking air density as 1.2 kg/m^3 (Chapter 10)
Air flow = 1094 m^3/h (0.30 m^3/s, 10.7 ft^3/s)

Cooling water flow

Assume a temperature rise of 10°C through the jacket and oil cooler. Then since the specific heat of water 4.18 kJ/kg°C and 1 kWh = 3600 kJ:

flow to jacket + oil cooler

$$\frac{125 \times 60}{4.18 \times 10} = 180 \ \text{kg/min} \ (180 \ 1/\text{min})$$

Exhaust flow

Sum of fuel flow + induction air flow = 1312.5 + 52.5 = 1365 kg/h

Energy balance for turbocharger[6]

In the case of turbocharged engines it is useful to separate the energy flows to and from the turbocharger and associated air cooler from those associated with the complete engine. In this way an energy balance may be drawn up covering the following:

- exhaust gas entering the turbine from engine cylinders
- induction air entering the compressor
- cooling water entering the air cooler
- exhaust leaving the turbine
- induction air leaving the cooler
- cooling water leaving the cooler.

A separate control volume contained within the control volume for the complete engine may be defined and an energy balance drawn up.

Summary

Calculation of energy balance from an engine test

(1)   Obtain information on the lower calorific value of the fuel.
(2)   From a knowledge of compression ratio of engine calculate air standard cycle efficiency as a yardstick of performance.
(3)   For one or a number of test points measure:

    (a)   fuel consumption rate;

    (b)   air flow rate to engine;

    (c)   exhaust temperature;

    (d)   power output;

    (e)   cooling water inlet and outlet temperatures and flow rate.

(4)   Calculate the various terms of the energy balance and the thermal efficiency from this data.

Prediction of energy balance for a given engine

(1)   Record type of engine and rated power output.
(2)   Calculate fuel flow rate from known or assumed specific fuel consumption.
(3)   Draw up energy balance using the guidelines given in Table 11.4.

Notation

lower calorific value	C_L MJ/kg
compression ratio	R
ratio of specific heats of air	$\gamma\,(=1.4)$

air standard cycle efficiency	η_{as}
mass flow rate of fuel	$\dot{m}_f$ kg/s
mass flow rate of inlet air	$\dot{m}_a$ kg/s
power output of engine	P_s kW
specific heat of air at constant pressure	C_p kJ/kg K
specific heat of water	C_w kJ/kg K
ambient temperature	T_a K
exhaust temperature	T_e K
cooling water inlet temperature	T_{1w} K
cooling water outlet temperature	T_{2w} K
mass flow rate of cooling water	$\dot{m}_w$ kg/s
For exhaust calorimeter:	
mass flow rate of cooling water	$\dot{m}_c$ kg/s
cooling water inlet temperature	T_{1c} K
cooling water outlet temperature	T_{2c} K
temperature of exhaust leaving calorimeter	T_{co} K

References

1. BS 7420 *Guide for Determination of Calorific Values of Solid, Liquid and Gaseous Fuels* (including Definitions).

2. *A.S.T.M. Special Technical Publication 525* (1972) *LP-Gas Engine Fuels*, American Society for Testing and Materials.

3. Goodyear, E.M. (1980) *Alternative Fuels*, Cranfield, Macmillan, London.

4. Heywood, J.B. (1988) *Internal Combustion Engine Fundamentals*, McGraw-Hill, Maidenhead.

5. Eastop, T.D. and McConkey, A. (1993) *Applied Thermodynamics for Engineering Technologists*, Longman, London.

6. Watson, N. and Janota, M.S. (1982) *Turbocharging the Internal Combustion Engine*, Wiley-Interscience, New York.

12 Measurement of mechanical losses in engines

This chapter is devoted to a critical study of the various methods, none of them totally satisfactory, of determining the mechanical efficiency of an internal combustion engine.

It is a curious fact that, in the long run, all the power developed by all the road vehicle engines in the world is dissipated as friction: either mechanical friction in the engine and transmission, rolling resistance between vehicle and road or wind resistance.

Mechanical efficiency, a measure of friction losses in the engine, is thus an important topic in engine development. It may exceed 80% at high power outputs but is generally lower and is of course zero when the engine is idling.

Under mixed driving conditions for a passenger vehicle between one third and one half of the power developed in the cylinders is dissipated either as mechanical friction in the engine, in driving the auxiliaries such as alternator and fan, or as pumping losses in the induction and exhaust tracts.

Since the improvement of mechanical efficiency is such an important goal to engine and lubricant manufacturers an accurate measure of mechanical losses is of prime importance. In fact the precise measurement of these losses is a particularly difficult problem, to which no completely satisfactory solution exists.

Fundamentals

The starting point in any investigation of mechanical losses should ideally be a precise knowledge of the power developed in the engine cylinder. This is represented by the indicator diagram, Fig. 12.1, which shows the relation between the pressure of the gas in the cylinder and the piston stroke or swept volume. For a 4-stroke engine account must be taken of both the positive area A_1, representing the work done on the piston during the compression and expansion strokes, and the negative area A_2, representing the 'pumping losses', the work performed by the piston in expelling the exhaust gases and drawing the fresh charge into the cylinder. (In the case of an engine fitted with a mechanical supercharger or an exhaust gas turbocharger there are additional exchanges of energy between the exhaust gas and the fresh charge, but this does

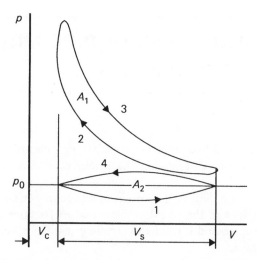

Figure 12.1.   *Indicator diagram, four-stroke engine. 1 induction; 2 compression; 3 expansion; 4 exhaust.*

not invalidate the indicator diagram as a measure of power developed in the cylinder.)

The universally accepted measure of indicated power is the indicated mean effective pressure (imep). This represents the mean positive pressure exerted on the piston during the working strokes after allowing for the negative pressure represented by the pumping losses.

The relation between imep and indicated power of the engine is given by the well-known expression:

$$P_i = \frac{\bar{p}_i V_s n}{60\,K} \times 10^{-1}\ \text{kW} \tag{1}$$

here $n/60\,K$ represents the number of power strokes per second in each cylinder.

The useful power output of the engine may be represented by the brake mean effective pressure

$$P = \frac{\bar{p}_b V_s n}{60\,K} \times 10^{-1}\ \text{kW} \tag{2}$$

The mechanical losses in the engine plus power to drive auxiliaries are represented by $(P_i - P)$ and may be represented by the friction mean effective pressure

$$\bar{p}_f = \bar{p}_i - \bar{p}_b \tag{3}$$

Mechanical losses from indicator diagram and measured power output

While this may appear to be the most logical method in practice it presents several difficulties:

- In a multi-cylinder engine indicator diagrams must be taken from all cylinders, so far as possible simultaneously. It is not safe to assume that the power developed in all cylinders is equal.
- There is a problem in obtaining an accurate record of the cylinder pressure, discussed in Chapter 20.
- Exact determination of true top dead centre position is difficult.

This last is a more serious difficulty than may be at first apparent. An electronic engine indicator records cylinder pressure in terms of crank angle, and measurements at each one degree of rotation are commonly available. However, in order to compute indicated power it is necessary to transform the cylinder pressure–crank angle data to a basis of cylinder pressure–piston stroke. This demands a very accurate determination of crank angle at the top dead centre position of the piston.

If the indicator records top dead centre 1° of crankshaft rotation ahead of the true position the computed imep will be up to 5% greater than its true value. If the indicator records tdc 1° late computed imep will be up to 5% less than the true value.

It is necessary to generate a signal precisely at top dead centre, usually by interaction between a pin in the rim of the flywheel and some kind of pickup. Precise determination of geometrical t.d.c. is not easy. The usual method is to rotate the engine to positions equally spaced on either side of tdc, using a dial gauge to set the piston height, and to bisect the distance between these points.

There are several sources of error:

- The difficulty of carrying out this operation with sufficient accuracy.
- The difficulty of ensuring that the signal from the pickup, when running at speed, coincides with the geometrical coincidence of pin and pickup.
- Torsional deflections of the crankshaft, which are always appreciable and are likely to result in discrepancies in the position of tdc when the engine is running, particularly at cylinders remote from the flywheel.
- It is not realistic, for all these reasons, to place much faith in measurements of indicated power based on the indicator diagram. Several other techniques, each having their limitations, will be described.

Motoring tests

A second method of estimating mechanical losses involves running the engine under stable temperature conditions and connected to a 4-quadrant dynamometer. Ignition or fuel injection are then cut and the quickest possible measurement made of the power necessary to motor the engine at the same speed.

Sources of error include:

- Under non-firing conditions the cylinder pressure is greatly reduced, with a consequent reduction in friction losses between piston rings, cylinder skirt and cylinder liner and in the running gear.
- The cylinder wall temperature falls very rapidly as soon as combustion ceases, with a consequent increase in viscous drag that may to some extent compensate for the above effect.
- Pumping losses are generally much changed in the absence of combustion.

Many detailed studies of engine friction under motored conditions have been reported in the literature, for a summary see ref. 1. These usually involved the progressive removal of various components: camshaft and valve train, oil and fuel pumps, water pump, generator, seals, etc. in order to determine the contribution made by each element.

The Morse test

In this test the engine is run under steady conditions and ignition or injection is cut off in each cylinder in turn. It is of course only applicable to multi-cylinder engines.

On cutting out a cylinder the dynamometer is rapidly adjusted to restore the engine speed and the reduction in power measured. This is assumed to be equal to the indicated power contributed by the non-firing cylinder. The process is repeated for all cylinders and the sum of the reductions in power is taken to be a measure of the indicated power of the engine.

A recent modification of the Morse test[2] makes use of electronically controlled unit injectors, allowing the cylinders to be disabled in different ways and at different frequencies, thus keeping temperatures and operating conditions as near as possible to normal.

The Morse test is subject, though to a less extent, to the sources of error described for the motoring test.

The Willan's line method

This is applicable only to unthrottled compression ignition engines. It is a matter of observation that a curve of fuel consumption rate against torque or bmep at constant speed plots quite accurately as a straight line up to about 75% of full power, Fig. 12.2. This suggests that for the straight line part of the characteristic equal increments of fuel produce equal increments of power; combustion efficiency is constant.

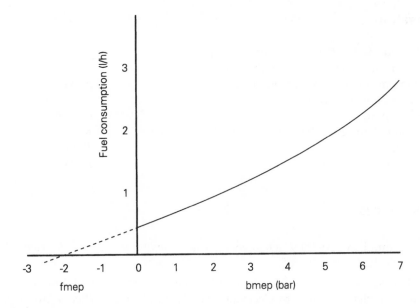

Figure 12.2. *Willan's line for a diesel engine*

At zero power output from the engine all the fuel burned is expended in overcoming the mechanical losses in the engine, and it is a reasonable inference that an extrapolation of the Willan's line to zero fuel consumption gives a measure of the friction losses in the engine.

Strictly speaking, the method only allows an estimate to be made of mechanical losses under no-load conditions. When developing power the losses in the engine will undoubtedly be greater.

Summary

The four standard methods of estimating mechanical losses in an engine and its auxiliaries have been described. No great accuracy can be claimed for any of these methods and it is instructive to apply as many of them as possible and compare the results. For a critical assessment see ref. 3. Measurement of mechanical losses in an engine is still something of an 'art'.

While no method can be claimed to give a precise absolute value for mechanical losses they are, of course, quite effective in monitoring the influence of specific changes made to a particular engine.

Notation

indicated mean effective pressure (imep)	$\bar{p}_i$	bar
brake mean effective pressure (bmep)	$\bar{p}_b$	bar
friction mean effective pressure (fmep)	$\bar{p}_f$	bar
swept volume of engine	V_s	1
engine speed	n	rev/min
constant, 1 for 2-stroke engines and 2 for 4-stroke	K	
indicated power	P_i	kW
engine power output	P	kW

References

1. Martin, F.A. (1985) *Friction in Internal Combustion Engine Bearings*, I. Mech. E. Paper C 67/85.
2. Haines, S.N.M. and Shields, S.A. (1989) The determination of diesel engine friction characteristics by electronic cylinder disablement, *Proc. I. Mech. E. Part A*, **203** (A2), 129–138.
3. Ciulli, E. (1993) A review of internal combustion engine losses. Part 2: Studies of global evaluations, *Proc. I. Mech. E. Part D*, **207** (D3), 229–240.

13 Effects of air/fuel ratio, fuel quality and mixture preparation – cylinder pressure diagrams and combustion analysis

This chapter deals with the process of combustion in the i.c. engine. Spark ignition and diesel combustion are described with an account of the effects of a number of different factors, including fuel properties. Cylinder pressure diagrams and the very important related topic of combustion analysis are discussed.

Fundamentals

In the last resort the performance of an internal combustion engine is largely determined by the events taking place in the combustion chamber and engine cylinder. These are influenced by a large number of factors:

- configuration of the combustion chamber and cylinder head
- flow pattern (swirl and turbulence) of charge entering the cylinder, in turn determined by design of induction tract and inlet valve size, shape, location, lift and timing
- ignition and injection timing, spark plug position and characteristics, injector and injection pump design, location of injector
- compression ratio (see Chapter 11)
- air/fuel ratio
- fuel properties
- mixture preparation
- exhaust gas recirculation (EGR)
- degree of cooling of chamber walls, piston and bore.

Combustion in the gasolene engine

An outline of the combustion process follows:

- A mixture of gas and vapour is formed, either in the induction tract, in which the fuel is introduced either by a carburetter or injector, or in the cylinder in the case of port injection. In every case the mixture enters the cylinder in what has been described as a 'shower of rain'.

- Combustion is initiated at one or more locations by an electric spark.
- After a delay, the control of which still presents problems, a flame is propagated through the combustible mixture at a rate determined, among other factors, by the air motion in the cylinder.
- Heat is released progressively with a consequent increase in temperature and pressure. During this process the bulk properties of the fluid change as it is transformed from a mixture of air and fuel to a volume of combustion products.

  Undesirable effects, such as pre-ignition, excessive rates of pressure rise, late burning and detonation or 'knock' may be present.
- Heat is transferred by radiation and convection to the surroundings.
- Mechanical work is performed by expansion of the products of combustion.

The development of the combustion chamber, inlet and exhaust passages and fuel supply system of a new engine involve a vast amount of experimental work, some on flow rigs that model the geometry of the engine, most of it on the test bed.

Great advances have been made in the study of induction, combustion and exhaust processes, involving much elaborate instrumentation, a full description of which would be outside the scope of this book. However a description of the effect of the most influential parameter, the air/fuel ratio, will help the test engineer unfamiliar with the subject to gain an overall understanding of the purpose of these techniques.

Effects of air/fuel ratio

Anyone who attempts to 'tune' a gasolene engine must have a clear understanding of this subject to make any progress with the task.

A gasolene engine is capable of operating on a range of air-to-fuel ratios by weight from about 8 : 1 to 20 : 1, and weaker than this in the case of stratified charge and lean-burn engines. Several definitions are important:

- Mixture strength. A loose term usually identified with air/fuel ratio and described as 'weak' (excess air) or 'rich' (excess fuel).
- Air/fuel ratio. Mass of air in charge to mass of fuel.
- Stoichiometric air/fuel ratio (sometimes known as 'correct' air/fuel ratio). The ratio at which there is exactly enough oxygen present for complete combustion of the fuel. Most gasolenes lie within the range 14 : 1 to 15 : 1 and a ratio of 14.5 : 1 may be used as a rule of thumb. Alcohols, which contain oxygen in their make-up, have a much lower value in the range 7 : 1 to 9 : 1.

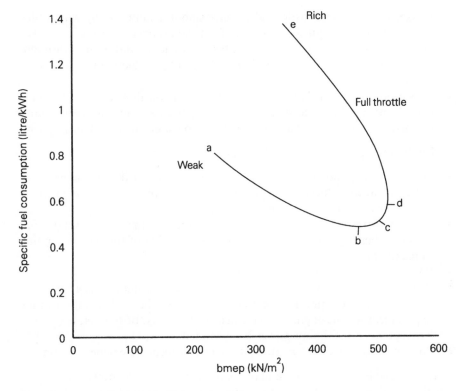

Figure 13.1. *Hook curve for a gasolene engine*

- Excess air factor or 'lambda' (Greek λ) ratio. The ratio of actual to stoichiometric air/fuel ratio. The range is from about 0.6 (rich) to 1.5 (weak). Lambda ratio has a great influence on power, fuel consumption and emissions.
- The equivalence ratio ϕ is the reciprocal of the λ ratio and is preferred by some authors.

A basic and very instructive test is to vary the air/fuel ratio over the whole range at which the engine is able to run, keeping throttle opening and speed constant. The results are often presented in the form of a 'hook curve', Fig. 13.1. which shows the relation between specific fuel consumption and bmep over the full range of mixture strengths.

If such a test is carried out on an engine fitted with a quartz observation window the following changes are observed:

- At mixture strength corresponding to maximum power and over a range of weaker mixtures combustion takes place smoothly and rapidly with a blue flame which is extinguished fairly early in the expansion stroke.

- With further weakening combustion becomes uneven and persists throughout the expansion stroke. 'Popping back' into the induction manifold may occur.
- As we proceed towards richer mixtures the combustion takes on a yellow colour, arising from incandescent carbon particles, and may persist until exhaust valve opening. This may lead to explosions in the exhaust system.

The following features of Fig.13.1 call for comment:

- Point *a* corresponds to the weakest mixture at which the engine will run. Power is much reduced and specific fuel consumption can be as much as twice that corresponding to best efficiency.
- Point *b* corresponds to the best performance of the engine (maximum thermal efficiency). The power output is about 95% of that corresponding to maximum power.
- Point *c* corresponds to the stoichiometric ratio.
- Point *d* gives maximum power, but the specific consumption is about 10% greater than at the point of best efficiency. It will be evident that a prime requirement for the engine management system must be to operate at point *b* except when maximum power is demanded.
- Point *e* is the maximum mixture strength at which the engine will run.

It is possible to produce similar curves for the whole range of throttle positions and speeds and hence to derive a complete map of optimum air and fuel flow rates as one of the bases for development of the engine management system, whether it be a traditional carburetter or a computer controlled injection system.

If at the same time that the hook curve is produced the air flow rate is measured the same information may be presented in the form of curves of power output and specific consumption against air/fuel ratio or λ, see Fig. 13.2, which corresponds to Fig. 13.1.

Effects of fuel quality

The most important properties of spark-ignition fuels are laid down in various specifications, see Table 13.1, (which does not cover many other, less critical, properties).

Octane number is the single most important gasolene specification, since it governs the onset of detonation or 'knock' in the engine. This condition, if allowed to continue, will rapidly destroy an engine. Knock limits the power, compression ratio and hence the fuel economy of an engine. Too low an octane number also gives rise to run-on when the engine is switched off.

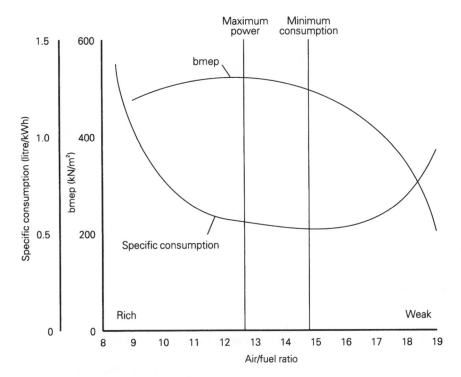

Figure 13.2. *Variation of power output and specific consumption with air/fuel ratio*

Three versions of the octane number are used: research octane number (RON), motor octane number (MON) and front end octane number (Rl00). RON is determined in a specially designed (American) research engine: the Co-operative Fuel Research (CFR) engine. In this engine the knock susceptibility of the fuel is graded by matching the performance with a mixture of reference fuels, iso-octane, RON = 100 and n-heptane, RON = 0.

The RON test conditions are now rather mild where modern engines are concerned and the MON test imposes more severe conditions but also uses the CFR engine. The difference between the RON and the MON for a given fuel is a measure of its 'sensitivity'. This can range from about 2 to 12, depending on the nature of the crude and the distillation process. In the UK it is usual to specify RON, while in the US the average of the RON and the (lower) MON is preferred. R100 is the RON determined for the fuel fraction boiling at below 100°C.

Volatility is the next most important property and is a compromise. Low volatility leads to low evaporative losses, better hot start, less vapour lock and

Table 13.1.   *Properties of gasolene: standards and test methods*

BSI	
BS 2000	Methods of test for petroleum and its products
ASTM	
D2623-83	Knock Characteristics of Liquefied Petroleum (LP) Gases by the Motor (LP) Method
D2699-84	Knock Characteristics of Motor Fuels by the Research Method
D2700-84	Knock Characteristics of Motor and Aviation-Type Fuels by the Motor Method
D2886-83	Knock Characteristics of Motor Fuels by the Distribution Octane Number (DON) Method
D2885-84	Research and Motor Method Octane Ratings Using On-Line Analyzers
D439-85a	Automotive Gasolene
D3710-83	Boiling Range Distribution of Gasolene and Gasolene Fractions by Gas Chromatography
D 240-85	Heat of Combustion of Liquid Hydrocarbon Fuels (General Bomb Method)
D2551-80	Vapor Pressure of Petroleum Products (Micromethod)
D 323-82	Vapor Pressure of Petroleum Products (Reid Method)
D2889-81	Vapor Pressures (True) of Petroleum Distillate Fuels
D 93-85	Flash Point by Pensky–Martens Closed Tester
Institute of Petroleum	
IP 34/80	Flash Point by Pensky–Martens Closed Tester
IP325/80	Front End Octane Number (RON 100°C) of Motor Gasolene
IP 12/73	Heat of Combustion of Liquid Hydrocarbon Fuels
IP236/69	Knock Characteristics of Motor and Aviation Type Fuels by the Motor Method
IP237/69	Knock Characteristics of Motor Fuels by the Research Method
IP 15/67	Pour Point of Petroleum Oils
IP171/65	Vapour Pressure – Micro Method
IP 69/78	Vapour Pressure – Reid Method

less carburetter icing. High volatility leads to better cold starting, faster warm-up and hence better short-trip economy, also to smoother acceleration.

Calorific value

A further very important fuel property that is often ignored is the calorific value. Engine development work is often concerned with 'chasing' very small improvements in fuel consumption, differences that can easily be swamped by variations in the calorific value of the fuels. Similarly comparisons of identical engines manufactured at different sites will be invalid if they are tested on different fuels.

For example, the LCVs of hexane and benzene, typical constituents of gasolene, are respectively 44.8 MJ/kg and 40.2 MJ/kg. A typical value for gasolene would be 43.9 MJ/kg, but this could easily vary by $\pm 2\%$ in different parts of the world, while the presence of alcohol as a constituent can depress the calorific value substantially.

Mixture preparation

Ideally, each cylinder of a multi-cylinder engine should receive under all operating conditions exactly the same charge of fuel and air, in exactly the same condition, as every other cylinder. This is an unattainable ideal, but a large element in the development programme of any new engine is concerned with approaching it as closely as possible.

A principal variable is the method of introducing the fuel. In order of increasing cost there are essentially three solutions: carburetter, single (throttle body) injector and individual port injectors. The main factors to be considered are:

- Equal distribution of air quantity, involving detail design of manifold and interaction with flow leaving throttle or carburetter. Particularly difficult at part throttle.
- Equal distribution of fuel. With carburetter or single-point injection there are three aspects to be considered: distribution of air, of fuel droplets and of fuel vapour. The proportion of the latter increases as power is reduced with accompanying increased manifold depression.
- Deposition of liquid fuel on manifold walls.
- Nature of mixture: degree of atomization and of vapour formation.

According to ref. 1 relative power output per litre swept volume for engines in the 1.3–1.6 litre range may be expected to lie in the following range:

Carburetter	41 kW/l
Throttle body injection	50 kW/l
Inlet port injection	53 kW/l

Combustion in the diesel engine

An outline of the combustion process follows:

- Air is drawn into the cylinder, without throttling, but frequently with pressure charging. Compression ratios range from about 14:1 to 22:1, depending on the degree of supercharge, resulting in compression pressures in the range 40–60 bar and temperatures from 700°C to 900°C.

- Fuel is injected at pressures that have increased in recent years from 600 bar to 1500 bar or higher. Charge air temperature is well above the auto-ignition point of the fuel. Fuel droplets vaporize, forming a combustible mixture which ignites after a delay which is a function of charge air pressure and temperature, droplet size and fuel ignition quality (cetane number).

- Fuel subsequently injected is ignited immediately, and the progress of combustion and pressure rise is to some extent controlled by the rate of injection. Air motion in the combustion chamber is organized to bring unburned air continually into the path of the fuel jet. Choice of combustion 'profile' involves a number of compromises. A high maximum pressure has a favourable effect on fuel consumption but increases NO_x emissions, while a reduction in maximum pressure, brought about by retarding the combustion process, results in increased particulate emissions. Maximum combustion pressures can exceed 200 bar.

Unlike the spark-ignition engine, the diesel must run with substantial excess air to limit the production of smoke and soot. A pre-combustion chamber engine can run with a lambda ratio of about 1.2 at maximum power while a direct injection (DI) engine requires a minimum excess air factor of about 50%, roughly the same as the maximum at which a spark-ignition engine will run. It is this characteristic that has led to the widespread use of pressure charging, which achieves a reasonable specific output by increasing the mass of the air charge.

Large industrial or marine engines invariably use direct injection; in the case of vehicle engines the indirect injection or pre-chamber engine is being abandoned in favour of direct injection because of the better fuel consumption and cold starting performance of the DI engine. Compact very high pressure injectors and electronic control of the injection process have made this development possible.

Properties of diesel fuels

Just as the range of size of the diesel engine is much greater than that of the spark-ignition engine, from 1–2 kW to 50 000 kW, the range of fuel quality is correspondingly great. Table 13.2 summarizes the main relevant specifications.

Cetane number is the most important diesel fuel specification. It is an indication of the extent of ignition delay: the higher the cetane number the shorter the ignition delay, the smoother the combustion and the cleaner the exhaust. Cold starting is also easier the higher the cetane number.

Viscosity covers an extremely wide range. BS 2869 specifies two grades of vehicle engine fuels, Class A1 and Class A2, having viscosities in the range

Table 13.2.   *Properties of diesel fuels: standards and test methods*

BSI

BS 2000.	Methods of test for petroleum and its products
BS 2869.	Fuel oils for oil engines and burners for non-marine use
BS MA 100.	Petroleum fuels for marine oil engines and boilers

ASTM

D 613-84	Ignition quality of diesel fuels by the cetane method
D 975-81	Diesel fuel oils
D 976-80	Cetane index, calculated, of distillate fuels
D 189-81	Carbon residue, Conradson, of petroleum products
D 524-81	Carbon residue, Ramsbottom, of petroleum products
D2500-81	Cloud point of petroleum oils
D 93-85	Flash point by Pensky–Martens closed tester
D 240-85	Heat of combustion of liquid hydrocarbon fuels (general bomb method)
D 97-85	Pour point of petroleum oils
D3245-85	Pumpability test for industrial fuel oils
D 129-64	Sulfur in petroleum products by the bomb method
D4294-83	Sulfur in petroleum products by nondispersive X-ray fluorescence spectrometry
D1552-83	Sulfur in petroleum products (high-temperature method)
D1266-80	Sulfur in petroleum products (lamp method)
D2709-82	Water and sediment in distillate fuels by centrifuge

Institute of Petroleum

IP218/67	Calculated cetane index of diesel fuels
IP 13/78	Conradson carbon residue of petroleum products
IP 34/80	Flash point by Pensky–Martens closed tester
IP 12/73	Heat of combustion of liquid hydrocarbon fuels
IP 41/60	Ignition quality of diesel fuels
IP 15/67	Pour point of petroleum oils
IP 14/65	Ramsbottom carbon residue of petroleum products
IP107/73	Sulphur in petroleum products

1.5–5.0 cSt and 1.5–5.5 cSt respectively at 40°C. BS MA 100 deals with fuels for marine engines. It specifies 9 grades, Class M1, equivalent to Class A1, with increasing viscosities up to Class M9, which has a viscosity of 130 cSt max at 80°C. These very heavy residual oils require elaborate methods of handling and purification.

Sulphur content is another major variable. Class A1 and Class A2 specify maximum sulphur contents of 0.3% and 0.5% respectively although current regulations, p.251, specify a maximum of 0.3% and this is to be further reduced to 0.05%. The permitted sulphur content of marine fuels ranges from 1%, Class Ml, up to 5% Class M6 and above. However, legislation limiting the permitted level of sulphur in marine fuels is under consideration.

Calorific value

The lower calorific value (net specific energy) of diesel fuels can range from roughly 40 MJ/kg to 43 MJ/kg and is a function of fuel density and sulphur content: the higher the density and the higher the sulphur content the lower the calorific value. Methods of calculating the approximate calorific value from density and sulphur content are given in refs. 2 and 3.

Cylinder pressure diagrams and combustion analysis

Many special techniques have been developed for the study of the combustion process. The oldest is direct observation of the process of flame propagation, using high-speed photography through a quartz window. More recent developments include the use of flame ionization detectors (FIDs) to monitor the passage of the flame and the proportion of the fuel burned, also hot-wire and laser doppler anemometry.

The most useful tool for the study of the combustion process is undoubtedly the cylinder pressure indicator, which has evolved from the simple 'cathode ray tube' of the 1930s to systems capable of examining the whole process in real time. The computerized investigation of combustion phenomena makes heavy demands on the data acquisition system. A variety of quantities must be measured, using appropriate transducers, either on a time base or synchronized with crankshaft position. They may include:

- fuel line pressure and needle lift
- cylinder pressure
- ionization signals
- crank angle
- time
- ignition system events
- inlet and exhaust pressures
- various (quasi-static) temperatures.

Each signal calls for individual treatment and appropriate recording methods, otherwise the computer memory may be swamped with largely redundant data.

Data acquisition rates have increased in recent years and systems are available with a sampling rate of up to 1 MHz on 16 channels. This is equivalent to cylinder pressures taken at 0.1° intervals at 16 000 rev/mm.

Combustion analysis aims at an understanding of all features of the process, in particular of the profile of heat release. It is perhaps the most powerful tool available for studying the thermodynamic processes that are central to the functioning of both spark-ignition and diesel engines. The test engineer concerned with engine development should be familiar with both the principles involved and the considerable problems of interpretation of results that arise.

A comprehensive account of the theory of combustion analysis would much exceed the scope of this book but a description of the essential features follows. The aim of combustion analysis is to produce a curve relating *mass fraction burned* (in the case of a spark ignition engine) or *cumulative heat release* (in the case of a diesel engine) to time or crank angle. Derived quantities of interest include rate of burning or heat release per degree crank angle and analysis of heat flows based on the first law of thermodynamics.

Stone and Green-Armytage[5] describes a simplified technique for deriving the burn rate curve from the indicator diagram using the classical method of Rassweiler and Withrow[4]. This takes as the starting point a consideration of the process of combustion in a constant volume bomb calorimeter. Figure 13.3 shows a curve of pressure and rate of change of pressure against time. It is then assumed that mass fraction burned and cumulative heat release at any stage of the process are directly proportional to the pressure rise.

Combustion in the engine is assumed to follow a similar course, the difference being that the process does not take place at constant volume. There are three different effects to be taken into account:

- pressure changes due to combustion
- pressure changes due to changes in volume
- pressure changes arising from heat transfer to or from the containing surfaces.

There has been much discussion of this problem in the literature[5,6] and no definitive solution. Stone's method will however be found adequate for comparative tests such as are usual in development work.

Figure 13.4 shows a pressure-crank angle diagram and indicates the procedure. The starting point is the 'no burn' curve, shown dotted. The most satisfactory way of obtaining this curve is undoubtedly to interrupt the ignition or injection for a single cycle and to record the corresponding pressure diagram; certain engine indicators are able to do this. The alternative, less accurate, method is to fit a polytropic compression line to the compression curve prior to the start of combustion and to extrapolate this on the assumption that the polytropic index n_c in the expression $pv^{n_c} = \text{constant}$ remains unchanged throughout the remainder of the compression stroke.

If A and B are two points on the compression line at which volumes and (absolute) pressures are respectively v_1, v_2 and p_1, p_2, then the index of compression is given by:

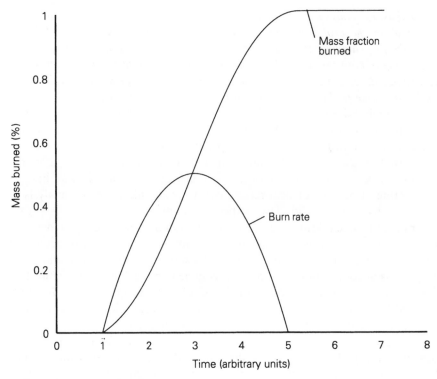

Figure 13.3.   *Combustion in constant volume bomb calorimeter*

$$n_c = \frac{\log (p_1/p_2)}{\log (v_2/v_1)} \tag{1}$$

The value of the index n_c during compression is likely to be in the region of 1.3, to be compared with the value of 1.4 for the adiabatic compression of air. The lower value is the result of heat losses to the cylinder walls and, in the case of spark ignition engines, to the heat absorbed by the vaporization of the fuel. The start of combustion is reasonably well defined by the point C at which the two curves start to diverge.

It is now necessary to determine the point D Fig. 13.4, at which combustion may be deemed to be complete. Various methods have been proposed, see refs. 5 and 6, but perhaps the most practical one is a variation on the method described above for determining the start of combustion. Choose two points on the expansion line sufficiently late in the stroke for it to be reasonable to assume that combustion is complete but before exhaust valve opening, 90° and 135° after t.d.c. may be a reasonable choice. The polytropic index of expansion (in the absence of combustion) n_e is calculated from eq. (1). This is also likely to be in the region of 1.3. (See ref. 6 for a discussion of the special case when

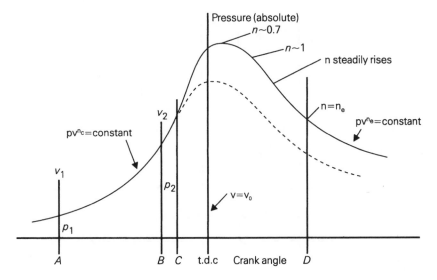

Figure 13.4.   *Pressure–crank angle diagram showing derivation of mass fraction burned*

burning continues right up to exhaust valve opening, as in a spark-ignition engine burning a weak mixture.)

The next step is to calculate the polytropic index of compression/expansion for successive intervals, typically one degree of crank angle, from the start of combustion. Once again eq. (1) is used, inserting the pressures and volumes at the beginning and end of each interval n varies widely as combustion proceeds but towards the end of the process. It converges on the value n_e determined above. The point D at which the two indices become equal is generally ill-defined, as there is likely to be considerable scatter in the values derived for successive intervals, but fortunately the shape of the burn rate curve is not very sensitive to the position chosen for point D.

The curve of heat release or cumulative mass fraction burned is derived as follows. Figure 13.5(a) shows diagrammatically an element of the indicator diagram before t.d.c. but after the start of combustion. During this interval the pressure rises from p_1 to p_2 while the pressure rise corresponding to the (no combustion) index n_c is from p_1 to p_0. It is then assumed that the difference between p_2 and p_0 represents the pressure rise due to combustion. It is given by:

$$\Delta p = p_2 - p_0 = p_2 - p_1 \left(\frac{v_1}{v_2} \right)^{n_c} \tag{2}$$

It is now necessary to refer this pressure rise to some constant volume, usually taken as v_0, the volume of the combustion chamber at t.d.c. It is assumed that the pressure rise is inversely proportional to the volume:

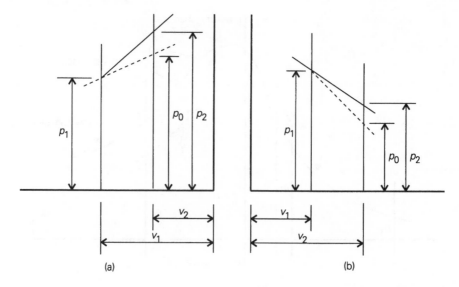

Figure 13.5.   *Derivation of mass fraction burned: (a) compression; (b) expansion*

$$\Delta p_c = \Delta p \frac{(v_1 + v_2)}{2v_0} \tag{3}$$

Essentially the same procedure is followed during the expansion process, Fig. 13.5(b), except that here there is a pressure fall corresponding to the (no combustion) index of expansion n_e. Δp and Δp_e are calculated from eqs (2) and (3) as before.

Finally Δp_c is summed for the whole combustion period and the resulting curve, Fig. 13.6, is taken to represent the relation between mass fraction burned or cumulative heat release and crank angle. The 'tail' of the curve, following the end of combustion, point D, will be horizontal if the correct value of n_e has been assumed. If n_e has been chosen too low it will slope downwards and if too high upwards.

There are a number of assumptions implicit in this method, for a clear discussion see ref. 6, but these do not seriously affect its value as a development tool. The technique is particularly valuable in the case of the DI diesel engine, in which it is found that the formation of exhaust emissions is very sensitive to the course of the combustion process, which may be controlled to some extent by changing injection characteristics and air motion in the cylinder. Figure 13.7, from ref. 7, shows the time scale of a typical diesel combustion process.

Figure 13.8 shows an indicator diagram taken from a small single-cylinder diesel engine and the corresponding heat release curve derived by means of a computer program modelled on the method of calculation described above.

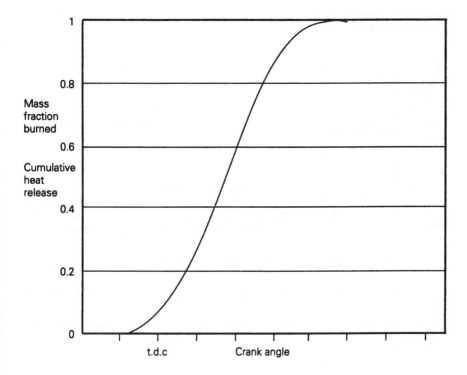

Figure 13.6. *Mass fraction burned*

The start of combustion, at 9° before t.d.c., is clearly defined and lags the start of injection by 6°. Combustion is complete by about 60° affer t.d.c. and the derived value of the (no combustion) index of expansion in this case was 1.36 (smaller engines generally display a higher index, owing to the higher heat losses in proportion to volume).

Evans[8], deals with the practical problems of choice of instrumentation and data logging systems. He emphasizes the need for extreme care in minimizing the effects of noise pick-up in the transducer leads.

The ability of present-day data acquisition systems to record and analyse a succession of combustion cycles in real time and to derive statistical parameters is of particular use in the development of the spark-ignition engine, where cycle-to-cycle variations in the combustion process have always been a problem, and are a major source of exhaust pollution. The way ahead appears to be by reducing variations in the ignition delay between the passage of the spark and the effective start of combustion. It is being overcome by attention to the small-scale turbulence or micro-turbulence in the combustion chamber.

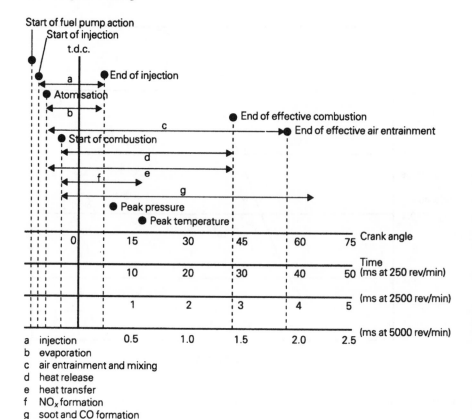

Figure 13.7.   *Time scale of diesel combustion process*

Summary

The processes taking place in the cylinder and combustion chamber are central to the performance of the internal combustion engine. In fact the rest of the engine may be regarded merely as a device for managing these processes and for extracting useful work from them.

The test engineer, even if mainly concerned with other aspects of engine performance, should be familiar with these processes, which are outlined in this chapter for spark ignition and diesel engines. The effects of air fuel ratio and fuel properties on performance are described. The technique of combustion analysis, which gives the clearest picture of in-cylinder processes, is described.

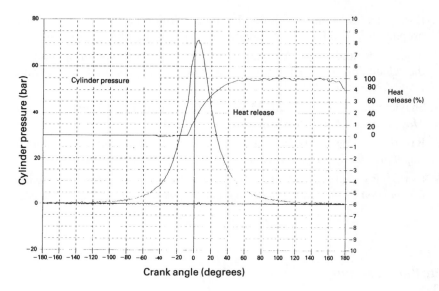

Figure 13.8. *Cylinder pressure diagram and heat release curve for a small single-cylinder diesel engine*

Notation

pressure, beginning of interval	p_1 bar
pressure, end of interval	p_2 bar
pressure, end of interval, no burn	p_0 bar
volume, beginning of interval	v_1 m^3
volume, end of interval	v_2 m^3
volume at top dead centre	v_0 m^3
pressure change due to combustion	Δp bar
pressure change, normalized to t.d.c.	Δp_c bar
polytropic index, general	n
polytropic index, compression	n_c
polytropic index, expansion (no burn)	n_e

References

1. Stone, R.D. *et al.* (1990) The design and development of the Rover K16 Engine, *Proc. I. Mech. E.*, **204**, Part D, 221–236.
2. BS MA 100 *Petroleum Fuels for Marine Oil Engines and Boilers.*
3. BS EN 590 *Specification for Automotive Diesel Fuel.*

4. Rassweiler, G.M. and Withrow, L. (1938) Motion pictures of engine flame propagation. Model for S.I. Engines, *S.A.E. Jl.*, **42**, 185–204 (Trans.)

5. Stone, C.R. and Green-Armytage, D.I. (1987) Comparison of methods for the calculation of mass fraction burnt from engine pressure-time diagrams, *Proc. I. Mech. E.*, **201** (D1).

6. Shayler, P.J. and Wiseman, M.W. (1990) *Improving the Determination of Mass Fraction Burnt*, S.A.E. Paper No. 900351.

7. Bazari, Z. (1992) *A D.I. Diesel Combustion and Emission Predictive Capability for Use in Cycle Simulation*, S.A.E. Paper No. 920462.

8. Evans, P.G. (1984) *A Fast Data-acquisition System for Investigating Combustion Phenomena in Spark-ignition Engines*, S.A.E. Paper No. 840057.

Further reading

BS 2869 Part 2 *Specification for Fuel Oil for Agricultural and Industrial Engines and Burners.*

BS 3016 *Specification for Pressure Regulators and Automatic Changeover Devices for Liquefied Petroleum Gases.*

BS 4040 *Specification for Leaded petrol (Gasolene) for Motor Vehicles.*

BS 4250 Part 1 *Specification for Commercial Butane and Propane.*

BS EN589 *Specification for Automotive LPG.*

BS 5355 *Specification for Filling Ratios and Developed Pressures for Liquefiable and Permanent Gases.*

BS 6843 Parts 0 to 3 *Classification of Petroleum Fuels.*

BS EN228 *Specification of Unleaded Petrol (Gasolene) for Motor Vehicles.*

BS 7405 *Guide to Selection and Application of Flowmeters for the Measurement of Fluid Flow in Closed Conduits.*

14 Exhaust emissions

It is probably true to say that at present a majority both of engine development and of routine testing is concerned with the impact of environmental legislation directed primarily towards the limitation and control of engine emissions.

This torrent of legislation, which shows no sign of diminishing, is particularly prolific in the field of vehicle emissions, but other areas, which include stationary engines, marine engines and the engines of railway locomotives, also receive increasing attention.

Even in the three years since the publication of the first edition of this book emissions regulations, particularly in the USA, have been made much more onerous. It is impossible therefore to summarize the current position with any confidence that this summary will continue to be valid, all that can be done is to give a 'snapshot' of the position as it is in 1998 and to indicate trends, so far as it is possible to foresee them.

One feature of the situation is of great practical importance: while the limits of permitted pollution, under pressure from a public obsessed with the subject, are moved inexorably downwards, it is not economically possible to change the test cycles that measure the pollution at the same rate. The cost and organizational complexity of developing and establishing a standard test procedure, even in a single country, are enormous and once a procedure is established it is essential that it should remain unchanged for as long a period as possible.

In this chapter the various harmful results of atmospheric pollution in general are described, and the contribution made by the internal combustion engine is assessed. This is followed by an account of the pattern of emissions from spark ignition and from diesel engines, and some indication of engine developments aimed at reducing these, including the various current methods of exhaust treatment.

A summary of current legislative requirements and established test procedures is given, including the specialized procedures for assessing evaporative emissions from vehicle fuel systems. Finally the range of instrumentation currently in use for the measurement of emissions and the design of emissions test cells are discussed.

Test cell design and operation: implications of emissions testing for the future

Test engineers not previously concerned with emissions testing should be aware of a number of special requirements that do not arise elsewhere in engine or vehicle testing:

- any trace of pollution of the combustion air by exhaust gas from other engines or fumes from industrial plant, e.g. paint, solvents, can upset the results of emissions tests. Cross contamination via ventilation ducts must be avoided and it may even be necessary to take into account the direction of prevailing winds in siting the cell.

- it will become more necessary to condition combustion air, calling for the kind of system described in Chapter 6, p.110 and requiring temperature control as a minimum.

- since much emissions testing will embrace cold starting and running at winter temperatures there will be an increased need for cells capable of operating at temperatures of $-10°$ C or lower.

- fuel treatment, see Chapter 6, p.93, will become considerably more demanding. It will be necessary to take great care in the production, storage and handling of fuel and in the avoidance of cross contamination and deterioration. Control of fuel temperature and instantaneous measurement will also become more exacting.

- the requirement to test vehicle engines with the full vehicle exhaust system will increase the average cell floor area and call for exhaust dilution and extraction ducts, see Chapter 6, p.112.

- the running of emission test cycles calls, in many cases, for expensive four quadrant dynamometers and elaborate control systems.

- a requirement will arise for testing onboard diagnostic (OBD) vehicle systems.

- the delicate and complex nature of much emissions measurement instrumentation, see p.256, will call for standards of maintenance, cleanliness and calibration at a level more often associated with medicine than with engineering.

Harmful consequences of atmospheric pollution

Effects of airborne lead

The harmful effects of lead, particularly on the health of children, have been known for many years and the pressure to eliminate the use of lead alkyls, a particularly poisonous form of the element, as an octane number improver for gasolenes has succeeded in effectively banning the production of leaded fuels in most parts of the world. A further incentive to the use of unleaded fuels is that a very small amount of lead will permanently 'poison' a catalytic exhaust converter.

Exhaust particulates

This is a subject of great complexity and rapidly increasing concern. A 1996 American paper[1] states:

The finest particulates, so-called 'soot cores', in diesel exhaust gas essentially consist of elementary carbon. The latest toxicological research suspects that these are carcinogens. The health hazards are due to the bio-incompatibility of the soot, probability of deposition in the alveoli, as well as the size ratio to the cell dimensions. All these deleterious factors point to the toxicity of the finest particulates in diesel exhaust gas.

Only the briefest summary of the current (1998) position will be attempted. The following definitions are important:

PM_{10}: particulate matter of aerodynamic diameter $< 10\,\mu m$ ($= 10$ micron $= 0.01\,mm$). Most airborne particles in the UK fall within this size range. Particles of this size can penetrate the human lung.

$PM_{2.5}$: particulate matter of aerodynamic diameter $< 2.5\,\mu m$, capable of penetrating the deep lung.

Exhaust Particulate Mass (PM): total mass collected on a standard filter over a specified drive cycle.

By far the largest *number* of exhaust particles emitted by both diesel and gasolene engines are of diameter < 100 nanometre ($= 0.1\,\mu m$), and medical research, such as that quoted above, indicates that particulates in this size range are particularly dangerous to health. The Motor Industry Research Association (MIRA) have made a special study of the various instruments available for measuring the distribution of number and size of exhaust particles, none of them suited to use outside a laboratory environment. As the permitted and achievable limits are pushed down the accuracy of control and measurement will have to keep pace. Figure 14.1(a) and (b), reproduced by permission of AEA Technology, show the numerical distribution of particle size in the exhaust from a heavy duty diesel engine and a spark ignition engine running on leaded fuel.

Particulates leaving the exhaust pipe of a vehicle engine undergo a complex series of chemical reactions when they are cooled and mixed with the atmosphere, hence legislative tests are based on apparatus (dilution tunnels, see p. 251) which attempt to reproduce these reactions so that samples resemble the composition and concentration experienced by humans 'on the sidewalk'.

Acid rain

Since the early 1970s acid rain, with its harmful effects on trees and other vegetation, has attracted increasing public attention, although the problem was recognized long ago: in 1872 the first British Alkali Inspector published a book

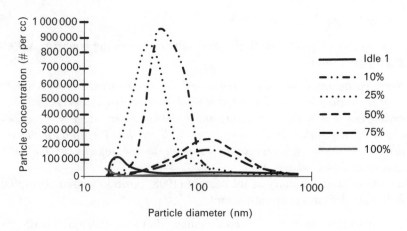

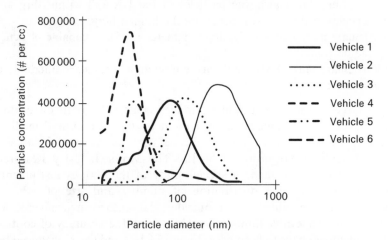

Figure 14.1.   *Particle size and concentration distribution (a) Euro 2: heavy duty diesel engine; (b) light duty gasolene engine at 100 kph*

on chemical meteorology in which he described rain from industrial areas as acid as any found today.

The mechanisms by which acid rain is formed involve chemical reactions that take place high in the atmosphere. The major acid precursors emitted as a result of man's activities are sulphur dioxide, SO_2, and nitric oxide, NO.

Sulphur dioxide undergoes oxidation to aerosol sulphate, HSO_3, in the presence of the hydroxyl radical, OH, itself produced by a complex chemical mechanism in the high atmosphere. The great majority of sulphur emissions arise from the combustion of coal and the quantity contributed by internal combustion engines is comparatively small, and diminishing under the pressure of legislation.

Table 14.1. *Concentration of pollutants in photo-chemical smog: typical values*

Pollutant	Concentration (Parts per 100 million by volume)
Carbon monoxide	3000
Ozone	25
Hydrocarbons	210
Sulphur dioxide	20
Oxides of nitrogen	20
Nitric acid	2

Nitric oxide is oxidized to nitrogen dioxide, NO_2, in the troposphere, mainly by reaction with ozone, itself formed by complex chemical processes involving carbon monoxide, hydrocarbons, water vapour and sunlight. Ozone itself, while its depletion in the stratosphere is a matter of concern, see below, has undesirable effects in the lower atmosphere.

Motor vehicles contribute roughly one third of the NO_x (a mixture of oxides of nitrogen) emissions in the UK plus carbon monoxide and unburned hydrocarbons. They are thus major contributors to the production of acid rain.

For a comprehensive review see ref. 2.

Photochemical smog[3]

Photochemical smog, first recognized in Los Angeles, is particularly apparent in cities where little coal is burned, there is little industrial activity and there are large concentrations of automobiles. Typical levels of the constituents of smog are shown in Table 14.1.

Global warming (the greenhouse effect)

This is another matter of general public concern, though much less well understood than the previous topics. It is well known that the existence of life on Earth depends on the ability of certain trace gases in the atmosphere to absorb and re-emit a substantial fraction of the infra-red radiation which the surface emits in response to solar heating and which in the absence of these gases would escape directly to space.

Water vapour is the most important natural greenhouse gas, but its concentration depends on the climate itself and is unaffected by human activities. The significant greenhouse gases that are affected by human activities

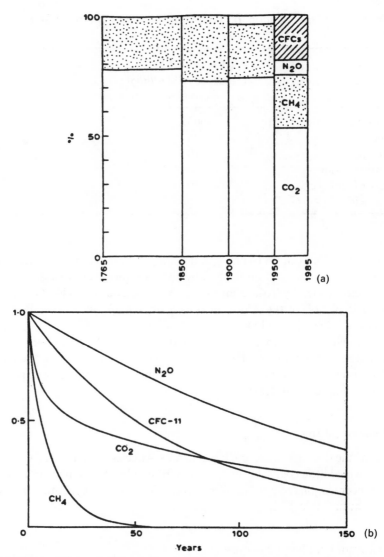

Figure 14.2. *Effect of different greenhouse gases: (a) relative contributions to the greenhouse effect from observed increases in atmospheric concentration of CO_2, methane, nitrous oxide and CFCs over different time periods; (b) fraction remaining in the atmosphere as a function of time*

are carbon dioxide, methane (CH_4), nitric oxide and CFCs (chlorofluorocarbons). Figure 14.2, reproduced from ref. 4, shows the relative contributions to the greenhouse effect of these different gases. The significance of CFCs, which have an atmospheric lifetime of the order of 100 years and an effect per unit mass of about 6500 times that of CO_2, is immediately apparent and explains the drastic steps taken (under an agreement known as the Montreal Protocol) to reduce the use of these substances. Methane emissions are largely the consequence of increasing population with cattle production and rice growing as the major man-made sources: the internal combustion engine is not a significant source.

Carbon dioxide is by far the most significant contributor and it is estimated that the CO_2 content of the atmosphere has increased from about 280 parts per million by volume (p.p.m.v.) in pre-industrial times to about 350 p.p.m.v. at present, with a current rate of increase of about 1.2 p.p.m.v. per annum. CO_2 emissions are more or less a direct function of fossil fuel consumption (the carbon/hydrogen ratio has a secondary influence), hence the pressure to increase engine efficiencies, the introduction, notably in the USA, of legislation prescribing vehicle fuel consumptions, the attempts to limit emissions on a global basis, and the various proposals for a so-called 'carbon tax'.

For a review of the subject see ref. 4.

Ozone depletion

The stratospheric ozone layer absorbs much of the Sun's ultra-violet radiation and any depletion of this ozone could have serious effects on human and animal life and on vegetation. CFCs represent the most serious threat to the ozone layer, but hydrocarbons and oxides of nitrogen are also harmful.

Atmospheric pollution: contributions by the internal combustion engine

It will be clear from the foregoing brief summary that the internal combustion engine is involved at many points in the problems of atmospheric pollution. The situation is summarized in Table 14.2.

Emissions from spark ignition engines

Perhaps the most characteristic feature of exhaust emissions, particularly in the case of the spark ignition engine, is that almost every step that can be devised in order to reduce the amount of any given pollutant has undesirable side effects, most frequently an increase in some other pollutant.

This is implied by some of the matters discussed in Chapter 13, notably Fig. 13.2, which shows the effect of variations in air/fuel ratio. Figure 14.3 is a development of this diagram, showing the effect on the emission of the main pollutants: CO, NO_x and unburned hydrocarbons. It will be apparent that one line of attack is to confine operation to a narrow window, around the stoichiometric ratio, and the main thrust of vehicle engine development in

Table 14.2.   *The internal combustion engine and atmospheric pollution*

Type of pollution	Principal sources	Relevance of the i.c. engine
Lead	Anti-knock compounds	A
Carcinogens	Diesel exhaust	A
Acid rain	Sulphur dioxide	B
	Oxides of nitrogen	A
	Unburned hydrocarbons	A
	Carbon monoxide	A
Global warming	CFCs	N
	Carbon dioxide	B
	Methane	N
Photochemical smog	Carbon monoxide	A
	Unburned hydrocarbons	A
	Sulphur dioxide	B
	Oxides of nitrogen	A
Ozone depletion	CFCs	N
	Unburned hydrocarbons	A
	Oxides of nitrogen	B

A: Major contributor
B: Secondary influence
N: Not involved

recent years has concentrated on the so-called 'stoichiometric' engine, used in conjunction with the three way exhaust catalyser, which converts these pollutants to CO_2 and nitrogen.

We have already seen, Fig 13.2, that a spark ignition engine develops maximum power with a rich mixture $\lambda \sim 0.9$, and best economy with a weak mixture, $\lambda \sim 1.1$ and before exhaust pollution became a matter of concern it was the aim of the engine designer to run as close as possible to the latter condition except when maximum power was demanded. In both conditions the emissions of CO, NO_x and unburned hydrocarbons are high: however the catalyser requires precise control of the mixture strength to within about ±5% of stoichiometric.

This requirement has been met by the introduction of fuel injection systems with elaborate arrangements to control mixture strength (lambda closed loop control). In addition detailed development of the inlet passages in the immediate neighbourhood of the inlet valve and the adoption of four-valve heads has been aimed at improving the homogeneity of the mixture entering the cylinder and developing small-scale turbulence: this improves the regularity of combustion, itself an important factor in reducing emissions. Other meas-

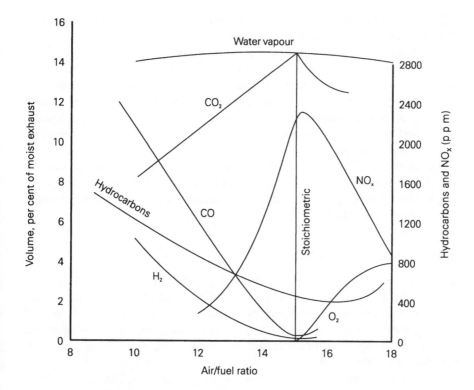

Figure 14.3. *Relation between exhaust emissions and air/fuel ratio, gasoline engine*

ures to improve the uniformity of the mixture include preheating of the intake air and steps to reduce the extent of the liquid fuel film on the intake passages.

Another area of development that significantly affects emissions concerns the ignition system. Spark plug design and location, duration and energy of spark, as well as the use of multiple spark plugs, all affect combustion and hence emissions, while spark timing has a powerful influence on fuel consumption as well as emissions. Here again conflicting influences come into play. Delaying ignition after that corresponding to best efficiency reduces the production of NO_x and unburned hydrocarbons, due to the continuation of combustion into the exhaust period, but increases fuel consumption.

Exhaust gas recirculation (EGR), involving the deflection of a proportion of the exhaust into the inlet manifold, reduces the NO_x level, mainly as a result of reduction in the maximum combustion temperature. The level of NO_x production is very sensitive to this temperature (this incidentally has discouraged development of the so-called adiabatic engine, in which fuel economy is improved by reducing in-cylinder heat losses). It also acts against the pursuit of higher efficiency by increasing compression ratios.

An alternative line of development is concerned with the *lean burn* engine, which operates at λ values of 1.4 or more, where NO_x values have fallen to an acceptable value, and fuel consumption is acceptable. However HC emissions are then high, the power output per unit swept volume is reduced and running under light load and idling conditions tends to be irregular. The engine depends for its performance on the development of a *stratified charge*, usually with in-cylinder fuel injection.

Development of exhaust after-treatment systems is a continuing process, at present particularly concerned with reducing the time taken for the catalyser to reach operating temperature, important because emissions are particularly severe during cold starts.

Emissions from diesel engines

The diesel engine presents rather different problems from the spark ignition engine. The diesel engine always operates with considerable excess air, so that CO emissions are not a significant problem, and the close control of air/fuel ratio, so significant in the control of gasolene engine emissions, is not required. On the other hand particulates are much more of a problem, and NO_x production is substantial.

The indirect injection (pre-chamber) engine performs well in terms of NO_x emissions; however the fuel consumption penalty associated with indirect injection has resulted in a general move to direct injection, associated with four-valve cylinder heads with a central fuel injector. This development is associated with a sharp increase in fuel injection pressures, now commonly 1500 bar or more, and the 'shaping' of the fuel injection pulse. The 'common rail' fuel injection system, where fuel is supplied at a uniform high pressure and injector needle movement controlled electrically, is making headway but carries a cost penalty.

NO_x emissions are very sensitive to maximum cylinder temperature and to the excess air factor. This has prompted the use of increased levels of turbocharging with improved after-cooling, also the use of retarded injection which results in reduced peak pressures and temperatures but, beyond a certain point, in increased fuel consumption. Exhaust gas recirculation can also play a part while, in large marine engines, the addition of up to 30% of water to heavy fuel oils can reduce emissions substantially.

Diesel particulate emissions, while greatly reduced since the days when heavy goods vehicles could be recognized by their black smoke, remain a problem. Developments in combustion chamber design and in fuel injection systems are helping, while attention to such factors as the reduction of lubricating oil consumption, the reduction of top land clearance and increased piston crown temperatures all make a contribution. However the evolution of effective exhaust after-treatment systems is not yet complete.

A further emissions problem, for which the engine cannot be blamed, concerns the presence of sulphur dioxide, SO_2, in the exhaust. Permitted levels of sulphur in fuels for road vehicles have been drastically reduced, see later, and have given rise to incidental problems with fuel injection equipment arising from the reduced lubricity of the fuel.

Current legislative requirements and test procedures

Vehicle engines

This is an immensely complicated and rapidly evolving field: all but the briefest summary would be impossible within the limits of this book, and would in any case be out of date almost as soon as it was published. The reader requiring detailed knowledge is advised to consult reference 5: *Air Pollution from Motor Vehicles: Standards and Technologies for Controlling Emissions*. World Bank (1996).

This publication summarizes International Standards originating from the USA, the U.N. Economic Commission for Europe (ECE) and the European Union (EU), together with national Standards for some 18 individual countries.

The majority of this legislation is concerned with specifying limits for the main pollutants: carbon monoxide, unburned hydrocarbons, oxides of nitrogen and exhaust particulates. This legislation has been largely driven by the USA, the first country to set emissions standards for vehicles, and the standards set by the US Environmental Protection Agency (EPA) have been adopted by many other countries. The generally less stringent standards set by the United Nations Economic Commission for Europe (ECE) are used in the European Union and a number of other countries. Japan has adopted rather different standards and test procedures.

Perhaps the best indication of the present state of progress can be given by summarizing some of the main American, ECE, and other standards. Table 14.3 indicates the current position and trends for the immediate future with regard to passenger car engines. In the case of all the four elements remarkable progress has been made. Table 14.4 shows the position for heavy duty (road) diesel engines, also the fairly exacting standards that are to be set for European railway locomotives. In the case of marine diesel engines, which in many cases are called upon to burn the heaviest and most polluted residual fuel, a start is being made in attempting to limit the NO_x emissions to a reasonable level.

The other main thrust of legislation has been to reduce the levels of lead and sulphur in fuels.

Table 14.3.   *Vehicle emissions standards: trends*

	Carbon monoxide	Hydrocarbons	Nitrogen oxides	Particulates (PM)
USA grams per mile, gasolene and diesel engines				
Light duty vehicles				
pre-1968 (uncontrolled)	90	15	6.2	
Tier 1, 1994–6	3.4	0.25	0.4	0.08
Tier 2, 2004	1.7	0.125	0.2	
Low emission vehicle (LEV) 10–14000 lb	7.0	0.3	2.0	
Ultra low emission vehicle (ULEV) 10–14000 lb	3.5	0.18	1.0	
Europe, grams per km				
Passenger cars, gasolene engines				
effective 2000	2.3	0.2	0.15	
effective 2005	1.0	0.1	0.08	
		$HC + NO_x$		
Passenger cars, diesel engines				
effective 2000	0.64	0.56		0.05
effective 2005	0.50	0.30		0.025

Table 14.4.   *Diesel emissions standards: trends*

	Carbon monoxide	Hydrocarbons	Nitrogen oxides	Particulates
USA grams per bhp-hr				
California, heavy duty	15.5	1.3	5.0	0.10
medium duty	14.4	$HC + NO_x$	3.5	0.10
Europe grams per kWh				
ECE 49 (13 mode)	14.0	3.5	18.0	(smoke meter)
Clean lorry directive				
Euro 3, 1999 (tentative)	2.5	0.7	5.0	<0.12
European Railways (UIC)	3.0	0.8	12.0	(Bosch 1.6–2.5)
Marine (draft IMO-regulation) —		—	10–17	—

Lead content of gasolene

In the U.S.A., Japan and many other countries the addition of lead to gasolene is banned. In Europe it is limited to 150 mg/l, all new cars must be capable of running on unleaded fuel, and a full ban will become effective in the year 2000. Worldwide, the share of the market held by leaded fuel is rapidly diminishing.

Sulphur content of diesel fuel

Diesel fuel for vehicle use normally contains from 0.1 to 0.5% sulphur by weight. SO_2 makes a significant contribution to particulate emissions and is recognized as a hazardous pollutant in its own right. Sulphur content is now limited to 0.05% in the U.S.A. and Europe, and a further reduction to 0.035% is expected by 2000. In the developing world sulphur contents tend to be much higher and an obstacle to reduction is the substantial cost of the necessary refinery modifications.

Emissions test procedures

The distinguishing feature of all emissions test procedures is that they all involve transient operation of the engine. They are thus vastly more demanding than any sort of steady-state test. They are, however, representative of the conditions of operation of vehicle engines.

A comprehensive survey of current and proposed test procedures would much exceed the scope of this book[5]. It is therefore proposed to describe the two principal current test procedures for passenger cars and to give a brief indication of the scope of other methods and anticipated developments.

The European exhaust emissions test procedure

The emissions test procedure for passenger cars was defined by ECE Regulation 15 and comprises the ECE 15 (urban) cycle and the EUDC (extra-urban) cycle. The ECE 15 cycle, Fig 14.4, is a simplified representation of the driving cycle in a typical European urban centre. It is followed immediately by the EUDC cycle, Fig. 14.5, which represents higher speed operation. There have been criticisms of these test procedures on account of their lack of severity, in particular of the modest acceleration rates, and it has been claimed that they under-estimate emissions by 15 to 25% compared with more realistic driving at the same speed[1].

The object of the test is to establish the mass of each nominated exhaust component emitted. The tests are carried out on a chassis dynamometer and consist of four repetitions of ECE15 followed immediately by EUDC. Initial heat soaking, start up and idling procedures are specified, but testing is only

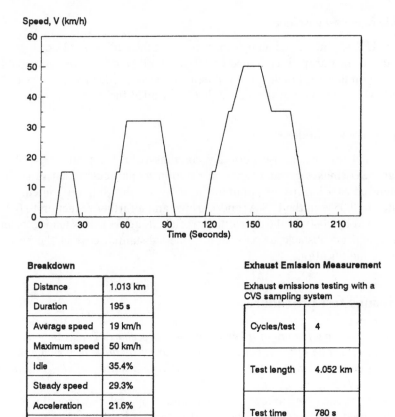

Figure 14.4.   *ECE 15 Emissions Test*

carried out between 20° and 30°C so low temperature CO emissions are not investigated.

The test installation is shown diagrammatically in Fig. 14.6, and uses the so-called constant volume sampling (CVS) technique with three collecting bags. Instrumentation is specified as:

CO emissions by non-dispersive infra-red analyser (NDIR)
hydrocarbons by flame ionisation detector (FID)
NO_x by chemiluminescence analyser
particulates by filtration

The US federal light duty exhaust emission test procedure (FTP-75)

This is a more complex procedure than the European test, and is claimed to more realistically represent actual road conditions. The cycle is illustrated in Fig. 14.7 and, in contrast to the European, embodies a very large number of

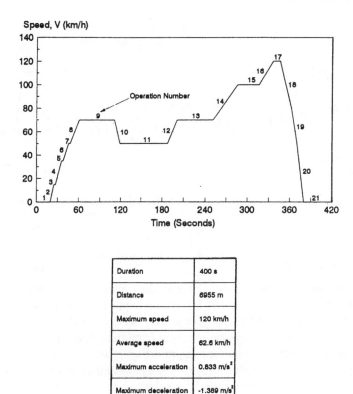

Duration	400 s
Distance	6955 m
Maximum speed	120 km/h
Average speed	62.6 km/h
Maximum acceleration	0.833 m/s²
Maximum deceleration	-1.389 m/s²

Figure 14.5. *EUDC (extra urban) Test Cycle*

speed changes. The test set up and instrumentation are essentially similar to that shown in Fig. 14.6. An additional sequence, the US06 cycle, has been introduced to cover the full range of speed and load conditions found in actual driving.

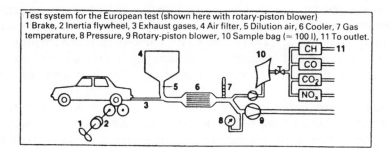

Figure 14.6. *Test installation for European exhaust emissions test procedure*

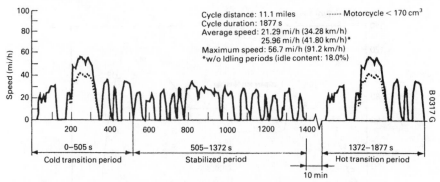

Figure 14.7.   *US federal light duty exhaust emission test procedure*

Other test procedures

The US procedure (1985) for testing heavy-duty engines prescribes a transient test cycle, while the European (ECE 49) and Japanese (13-mode) procedures use steady state tests. All three procedures, which are described in ref. 5, measure emissions with the engine removed from the vehicle and installed on a dynamometer. Results are reported in gram/bhp-hr or kW-hr.

The emissions test procedures for motor cycles and mopeds in force in the USA and in Europe are essentially similar to those described above for passenger cars. Testing is carried out on a single roll dynamometer with inertia simulation, with a clamp to hold the motorcycle upright.

Inspection and maintenance (I/M) test procedures

These tests, which (1998) are under active development will enormously extend the practice of emissions testing. They are intended to monitor the emissions aspect of performance of vehicles in service and will be carried out either in high-volume test facilities licensed by government or by private garages; the former are reputed to be more effective. In the USA the IM 240 procedure, which is based on the first 240 s of the FTP-75 procedure, uses a simple rolling road dynamometer and constant volume sampling to measure emissions per km over a realistic driving cycle. Annual tests in accordance with this procedure are to become mandatory.

In Europe there have been several false starts, for example the unfortunate Ministry of Transport procedure in the UK, which involved the free acceleration of diesel engines, with sometimes disastrous results. It seems likely that here, again, European practice will follow the American, with a consequent huge increase in demand for test facilities.

Evaporative emissions testing

The significance of evaporative emissions in causing pollution of the atmosphere by volatile organic compounds has only slowly been recognized. Gasolene is a relatively volatile fuel, and while this volatility is commonly adjusted to current conditions, losses from a vented fuel tank can be substantial.

The main sources of evaporative emissions are daily emissions, the consequence of daily temperature cycles, running losses, and hot soak losses. They are generally controlled by venting the fuel tank through a canister of activated charcoal, which absorbs the vapour and is regenerated by air drawn through when the engine is running.

The sealed housing evaporative determination (SHED) test procedure was evolved in the USA and has recently been made more severe in the light of the realization that emissions from this source were greater than had been appreciated. The *diurnal* evaporative losses are measured by placing the vehicle in the enclosure and measuring the emissions as the temperature in the fuel tank is increased, thus simulating the temperature rise during a day. In the modified procedure this process is repeated three times over a period of 72 hours.

The *running losses* are determined either by running the vehicle on a rolling road with absorbent carbon canisters attached to the various possible sources of emissions, or, in the latest procedure, running through three standard driving cycles in the SHED. The *hot soak* test measures the emissions for one hour immediately following the running test. Acceptable losses from the complete procedure are 2.0 gm of fuel per test, effectively zero.

SHED chambers, Fig. 14.8, are constructed of aluminium alloy or stainless steel and include elaborate arrangements for sealing and for maintaining the pressure difference between chamber and exterior to the close limit of ± 25 mm H_2O (either by the use of flexible panels or a floating roof).

Many countries have adopted variations of these test procedures. For regular updating see the annual report prepared by the CONCAWE organisation[6].

Instrumentation for the measurement of exhaust emissions

The great majority of engine test work is concerned with the taking of measurements that are in principle quite simple, even though great skill may be needed to ensure the right answer: forces, pressures, masses, flow rates, speeds, displacements, temperatures, oscillations.

Where emissions measurements are concerned we are forced to move into a totally different and very sophisticated field of instrumentation engineering. The apparatus makes use of subtle and difficult techniques borrowed from the field of physics.

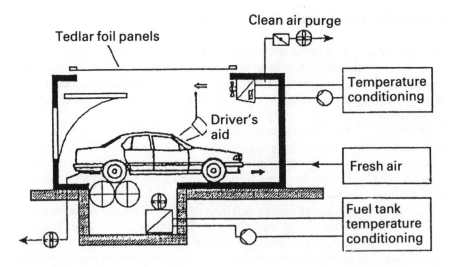

Figure 14.8.   *SHED (sealed housing evaporative determination) test cell with chassis dynamometer*

It is very desirable, for all but the simplest garage-type emissions apparatus, that a technician should be specially trained to take responsibility for the maintenance and calibration of the instruments involved.

Table 14.5, which is not exhaustive, lists the main types of instrument used

Table 14.5.   *Techniques of emissions analysis*

Gas	Technique	Typical range	90% response time
CO	Non-dispersive	0–3000 ppm	2–5 s
CO_2	infra-red, NDIR	0–20%	
NO_x	Chemiluminescence detector, CLD	0–1%	1.5–2 s
	Fast response CLD	0–1%	4 ms
Unburned hydrocarbons, UHC	Flame ionisation detector, FID	0–1%	1–2 s
	Fast FID	0–1%	1–2 ms
Hydrocarbon species, CH_4, SO_2, acids, etc.	Fourier transform infra-red, FTIR	various	5–15 s
Exhaust gas: variable flow rate	Helium trace flow meter, HTFR		1–2 s

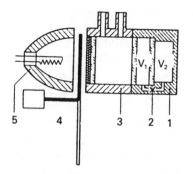

1 Receiver chamber with compensation spaces V_1 and V_2, 2 Flow sensor, 3 Measuring cell, 4 Rotating chopper disc with motor, 5 Infrared emitter

Figure 14.9.   *Measuring chamber of non-dispersive infra-red analyser*

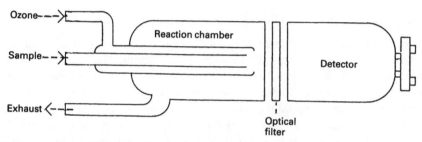

Figure 14.10.   *Chemiluminescence detector*

for measuring the various gaseous components of exhaust emissions. The measurement of particulate emissions is dealt with later, p.260. Continuing developments are taking place in this field, notably in the evolution of instruments having a more rapid response time.

Non-dispersive infra-red analyser, NDIR

This instrument, Fig. 14.9, makes use of the property of certain gases – of which carbon monoxide and carbon dioxide are particularly pronounced examples – of selectively absorbing infra-red radiation over a narrow range of wavelengths. The gas to be examined is drawn through a measuring cell through which pulses of broad-band infra-red light are passed. The degree of absorption of energy in the relevant wave-band is measured by a system of detector cells and gives a measure of the concentration of the corresponding gas in the sample.

Chemiluminescence detector, CLD

In this instrument, Fig. 14.10, use is made of the chemical reaction between nitric oxide and ozone:

$$NO + O_3 \rightarrow NO_2 + O_2 \rightarrow NO_2 + O_2 + photon$$

The nitrogen compounds in exhaust gas are a mixture of NO and NO_2, described as NO_x. In the detector the NO_2 is first catalytically converted to NO and the sample is reacted with ozone, generated by an electrical discharge through oxygen, at low pressure in a heated vacuum chamber. The light is measured by a photomultiplier and indicates the NO_x concentration in the sample.

A recent development, the Cambustion Fast Response CLD, makes use of a miniaturised reaction chamber and has a sufficiently rapid response time to permit the examination of cycle-by-cycle variation in NO_x emissions.

Flame ionisation detector, FID

The operation of this instrument, Fig. 14.11, depends on the production of free electrons and positive ions that takes place during the combustion of hydrocarbons. If the combustion is arranged to take place in an electric field the current flow between anode and cathode is closely proportional to the number of carbon atoms taking part in the reaction. The corresponding UHC level is thus to some extent affected by the hydrogen/carbon ratio of the fuel. In the detector the sample is mixed with hydrogen and helium and burned in a chamber which is heated to prevent condensation of the water vapour formed.

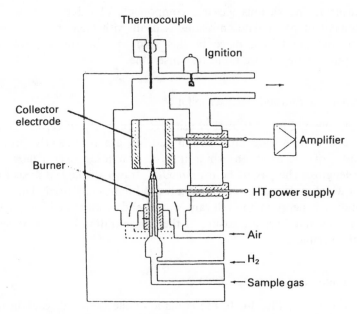

Figure 4.11.   *Flame ionization detector*

Fast FID

This is a miniaturized development of the previous instrument which is capable of a response time measured in milliseconds and may thus be used for in-cylinder and exhaust port measurements.

Fourier transform infra-red analyser, FTIR

This operates on the same principle as the NDIR, but performs a Fourier analysis of the complete infra-red absorption spectrum of the gas sample. This permits the measurement of the content of a large number of different components. The method is particularly useful for dealing with emissions from engines burning alcohol-based fuels, since methanol and formaldehyde may be detected.

Helium trace flow meter, HTFR

This recent development, by Horiba, depends for its operation on the injection of a small flow, $< 500 \, cc/min$, of helium into the exhaust flow, which may lie within the range $50-3000 \, l/min$. At a distance of not less than 1.2 m down-stream, to ensure full mixing of the helium, a sample is taken and the helium concentration measured, giving a direct indication of exhaust gas mass flow rate.

The 90% response time of less than 3 s means that the instrument may be used to plot the rapid changes in air flow rate through the engine associated with load and speed changes and turbocharger response.

All these methods, with the exception of the last, may be used either with samples drawn directly from the exhaust system or with the constant volume sampling, CVS technique, see p.252.

Response times

All these instruments are required to operate under one of two quite different conditions, depending on whether the demand is for an accurate measurement of a sample collected over a fairly long time interval, as in the case of the various statutory test procedures, or for an instantaneous measurement made under rapidly changing conditions, such as arise in true transient testing, the purpose of which is to study the performance of the engine and its control system in detail.

These two sets of requirements are conflicting: analysers for steady state work must be accurate, sensitive and stable and thus tend to have slow response times and to be well damped. Analysers for transient work do not require such a high standard of accuracy but must respond very quickly, preferably within a few milliseconds. Reduction of response time is a prime object of instrument development in this field.

A further problem in transient testing is concerned with the time and distance lags associated with the positioning of the exhaust gas sampling points. This matter is discussed in ref. 7, which puts forward a mathematical technique, of very general applicability, for reconstructing the true signal from the instrument signal, taking into account sampling delays and instrument response characteristics.

Particulate emissions

There are essentially three methods in use for measuring particulate emissions, apart from the laboratory-type apparatus for studying particle size distribution, p.241, and these methods cannot readily be related one to the other:

- the 'traditional' method, by the use of a smoke-meter which measures the opacity of the undiluted exhaust by the degree of obscuration of a light beam

- measurement of the particulate content of an undiluted sample of exhaust gas by drawing it through a filter paper of specified properties and estimating the consequent blackening of the paper against an agreed scale

- measurement of the actual mass of particulates trapped by a filter paper during the passage of a specified volume of diluted exhaust gas.

The Hartridge Smokemeter*

This instrument, shown diagrammatically in Fig. 14.12, has become effectively the European standard. A probe having a bore in the range 10–25 mm, depending on the size of engine, leads a sample of exhaust to the centre of a heated smoke tube. It flows towards each end of the tube where clean air, supplied by a fan, directs it into a sleeve surrounding the smoke tube and thence to discharge. Light from a halogen lamp is directed through the smoke tube and the light not absorbed by the smoke is detected by a silicon photodiode.

Smoke density is reported in terms of a coefficient of absorption k^* or a Hartridge smoke unit (HSU) having a range 0–100. Smoke meters are used either for static or dynamic (acceleration) tests in the routine testing of vehicles in service. Current European regulations are laid down in document EEC 24 but, like so much legislation in the field of emissions, are at the time of writing in a state of flux.

---

* None of these smokemeters is usable for the detection and diagnosis of white exhaust smoke

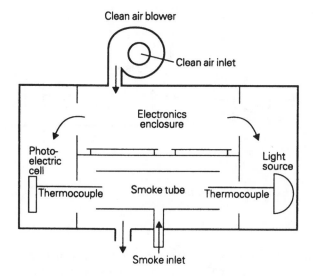

Figure 14.12.   *Hartridge Smokemeter*

The Celesco smokemeter

In this instrument the light beam traverses the exhaust duct at right angles to the flow, permitting rapid response to transient conditions. The photodiode signal is updated every 1.6 ms.

The Bosch filter-type smokemeter

This instrument, also effectively standardized in Europe, operates by drawing a fixed volume of exhaust gas (330 c.c.) through a filter paper under specified conditions of engine operation. The degree of blackening of the paper is assessed either by photovoltaic cell or visually against a reference scale graded in soot numbers of 0 to 9.

Exhaust gas dilution tunnels

Clearly the above methods give no indication, except by inference based on experience, of the actual mass of particulates present in a given volume of exhaust. For such measurements a very much more elaborate technique, suitable only for laboratory use, is necessary. This is associated with the constant volume sampling (CVS) technique Fig. 14.4. Again test procedures are in a state of flux, but the principles are clear. A sample of gas is drawn slowly through a filter paper in the course of a test sequence such as the ECE 15 cycle or the American FTP sequence.

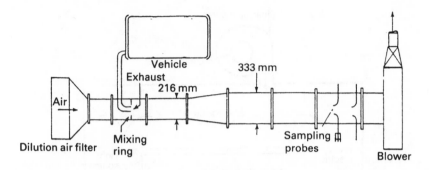

Figure 14.13. *Exhaust gas dilution tunnel (SAE J 1280 Dynamic Dilution System B)*

The special factor of most significance here is that it is laid down that the exhaust should be diluted with fresh air so as to simulate the situation of a vehicle in motion. The make up of a smoke particle in the exhaust pipe differs from that of the same particle after it has entered the atmosphere. This is because chemical reactions take place between particle and atmosphere: since it is the properties of the latter that are of interest the purpose of the test set-up is to imitate as closely as possible the real world situation. This is achieved by drawing air through a polished and heated stainless steel tube, typically 25 or 30 cm (10 or 12 in) in diameter at a velocity in the region of 10 m/s, Fig. 14.13. The exhaust from the engine enters the upstream end of the tube (the particulate tunnel) and turbulent mixing takes place.

A small sample is drawn from the downstream end of the tunnel and subjected to further dilution in a secondary tunnel. Finally a sample of gas from this tunnel is drawn through two filters of Teflon-coated glass fibre in series. The filter papers are weighed on a very high precision balance and, since the volumes of dilution air and of the sample passing through the filters are carefully controlled and measured, the total mass of particulates emitted in the course of the test run may be calculated. It is reported in g/km or for stationary engines in g/kWh.

Particularly for heavy vehicle engines the full flow particulate tunnel is a very bulky and expensive device. The so-called mini-dilution tunnel has been developed to meet this problem. Typically the miniature tunnel has a diameter of 25 mm (1 in) and a length of 635 mm (25 in) and handles only a small fraction of the exhaust flow. It is clearly necessary in this case to measure precisely the proportion (typically 2%) of the total exhaust flow entering the tunnel. This is achieved by the use of a specially designed sampling probe that is regulated to ensure that the velocity of the gas sample is precisely the same as the velocity in the exhaust pipe at the point of withdrawal.

Special features of emission test cells

A test cell intended for carrying out the statutory emissions test procedures will be very much larger than a simple engine test cell, since it must accommodate:

- the complete vehicle mounted on a rolling road dynamometer
- for diesel tests, the particulate tunnel which, with its associated equipment, may occupy a space of 1.5 m × 6 m
- various large control and instrumentation cabinets.

Equipment in the control room includes the control and measurement electronics plus gas handling equipment and instrumentation. Six or more standard 1.8 m (6 ft) cabinets may be required and the heat generated must be considered in designing the ventilation system. If the emission tests to be run include torque reversals a four-quadrant dynamometer will be necessary, with associated control cabinets which produce substantial heat and are best installed externally (the test cell roof may be a good location). The high cost of heavy current cabling should be kept in mind when planning the layout.

Finally, a store for reference gas bottles will be necessary. The emissions measuring equipment for gasolene engines differs considerably from that required for diesels and it may be sensible to have a separate bed for each type.

Summary

This is a very large subject, the importance of which can only increase. In this chapter the various harmful effects on the environment that may be laid at the door of the internal combustion engine are described and the current state of legislative requirements, which are becoming rapidly more exacting, is summarized. These are being translated into complex and expensive test procedures which are likely to occupy a growing fraction of all engine test facilities in years to come.

The main design features affecting the tendency of an engine to produce harmful emissions are described, together with an account of current attempts to improve performance in this respect. A description of the main types of emissions-measuring instrumentation and of the special features of emission test cells follows.

References

1. SAE 960472 (1996) Mayer, A. *et al.* Trapping efficiency depending on particulate size. SAE Congress Feb. 1996. SP 1140.

2. Report No.14 (1983) *Acid Rain Papers presented at the Fifteenth Consultative Council Meeting of the Watt Committee on Energy*, Watt Committee on Energy, London.
3. *Encyclopaedia of Science and Technology*, 6th edn. (1987), *Smog* Vol.16; pp. 474–476, McGraw-Hill, N.Y.
4. Report No. 23 (1990), *Technological Responses to the Greenhouse Effect.* Watt Committee on Energy, London.
5. Faiz, Asif *et al.* (1996) *Air Pollution from Motor Vehicles Standards and Technologies for Controlling Emissions,* World Bank, Washington DC
6. *Oil Companies European Organisation for Environmental Health and Protection* (CONCAWE) Madouplein 1, 81030 Brussels, Belgium.
7. Beaumont, A.J. *et al.* (1990) *Signal Reconstruction Techniques for Improved Measurement of Transient Emissions,* S.A.E. Technical Paper No. 900233.

Further reading

Horiba (1990) *Fundamentals of Exhaust Emissions Analysis and their Application,* Horiba Instruments Ltd, Northampton.
BS 1747 Parts 1 to 13 *Methods for Measurement of Air Pollution.*
BS 4314 Part I *Infra-red Gas Analysers for Industrial Use.*
BS 5849 *Method of Expression of Performance of Air Quality Infra-red Analysers.*

Useful addresses

National Environment Research Council, Polaris House, North Star Avenue, Swindon, Wilts, SN2 1EU.
Environmental Protection Agency, 401 M Street SW, Washington DC 20460, USA.

15   Transient testing

Transient testing is in practice almost confined to automotive engines, since industrial and marine engines are not subject to the almost continuous variations in load and speed that are characteristic of the automotive engine environment. Up to the mid-1990's the majority of transient engine testing was associated with emissions testing carried out in accordance with the kind of test procedure described in Chapter 14. Since that time the work associated with engine and drive train calibration, see below, has greatly increased, as has the sophistication of the models, which may be required to simulate both the road load and the vehicle power train.

In this Chapter transients are classified in terms of four characteristic time scales:

0–0.2 seconds, 0.5–2 seconds, 1.0–10 seconds, 1–10 minutes

It is shown that these classifications embrace a vast number of road vehicle control phenomena, requiring a large number of different experimental techniques for their examination. The road load equation, the essential theoretical basis for much transient testing, is derived and illustrated by an example.

Engine calibration and homologation are two specialized areas of engineering activity which increasingly use transient testing as a tool and are thus appropriately treated in this Chapter.

Engine and drive train calibration

The word 'calibration', when used in this connection, does not have the meaning usually attributed to it. The term originally meant 'the process of determining the calibre of a gun', but has long been extended to include the process of verifying the scale of an instrument and determining the extent of any errors.

In the present case 'calibration' means the precise measurement and optimization of the characteristics of an engine and drive train in all possible vehicle model variants, the basic engine (crankcase and running gear) being a constant but many other features being variable.

It costs a vast amount of money to develop a new engine and the enterprise is only economically viable if the engine then finds the widest possible field of application. Today this field is essentially world-wide, and the developers will hope that their engine will be incorporated in a very wide range of vehicles, manufactured in many different countries, each having their own customer preferences and legislative requirements.

Clearly the matching of engine and drive train to the requirements of a particular customer must be a two-way process of successive approximation, whereby the engine and drive train are gradually brought into line with what is needed for the particular application. Equally obviously the process cannot sensibly be carried out 'in a vacuum': as much data as possible regarding the range of performance of the engine and drive train that is on offer must be available.

A further important purpose of the calibration procedure is the reduction of development time: the sooner the matrix of testing needed to establish all the parameters of the new product is completed the sooner its fitness for the market can be established and the more rapid the process of matching it to the requirements of individual customers can be.

Homologation

This is another case of a word, in this case having originally no particular engineering associations, being taken over and given a specialized meaning. The Oxford English Dictionary defines 'to homologate' as 'to express agreement with, approve, acknowledge, confirm, ratify'.

Knowledge of and compliance with the legislative requirements of different markets, covering such areas as safety, exhaust emissions, fuel consumption rates, noise vibration and harshness, competitive benchmarking, etc. is an equally important aspect of the problem of gaining acceptance of a given product, in this case an assembly of components forming a vehicle, in all the various markets in which it is to be sold.

For the engine or vehicle manufacturer this is another complex and expensive area of activity. All major versions, and all derivatives, of the vehicle must meet the formal requirements, some of them having legislative backing, that are in force in each country in which the vehicle is to be sold. Homologation is the process of establishing this conformity, both for whole vehicles and for components.

Benchmarking

This is merely a modern term for an activity that has been practiced by makers of products intended for sale, probably ever since the first maker of flint axes went into business: it is the act of comparing your product with competing products that are also on offer and your production and testing methods with those of your competitors. The difference today is that it is now highly formalized, and practised without compunction. Once it is on the market any vehicle or component thereof can be bought and tested by the manufacturer's competitors, with a view to taking over and copying any features that are clearly in advance of the competitor's own products.

This evidently increases the importance of patent cover, of preventing the transfer of confidential information by disaffected employees, and of maintaining confidentiality during the development process. This latter is a matter of great concern to engine development establishments that work for a number of clients.

The necessity for transient testing: the problem

Any vehicle user is aware that the performance of an engine is to a considerable extent judged by its characteristics when changing state. A cloud of black smoke from a truck pulling away from traffic lights, or hesitant acceleration when overtaking are just two of many examples of inadequate performance that is a direct consequence of unsatisfactory transient behaviour on the part of the engine.

It is perhaps worth pointing out that while in what may be described as 'classical' steady state testing the transitions between successive test conditions are unimportant, in transient testing the transients *are* the test sequence.

The test engineer involved in transient testing is faced with two challenges, first to ensure that the test sequence accurately models the conditions experienced by the engine in service and secondly to ensure that the sequence is precisely repeatable.

It is simply not practicable to carry out detailed transient testing, which must include analysis of engine and vehicle response, on the road, although road tests are, of course, the final (subjective) means of assessing the actual handling of the vehicle. The favoured modern solution is the use of a direct coupled four-quadrant a.c. dynamometer with computer control; such a machine can be designed to react sufficiently rapidly to simulate all the transient conditions experienced by the engine on the road.

The classification of transient conditions

The term transient testing covers a very wide range of investigations. These may be roughly classified in terms of the characteristic time scales involved.

Time scale 0–0.2 sec, frequencies 5–500 Hz

In this range we are concerned essentially with torsional vibrations in the engine–transmission–road wheel complex ranging from two–mass engine–vehicle judder, with a frequency typically in the range 5–10 Hz, to much higher frequency oscillations involving the various components of the power train. Subtle effects can occur as a result of interaction with the engine mountings. Rotation of the engine on its mountings not only affects the dynamics of the whole system but may disturb the throttle linkage, with consequent variations in the throttle position.

This kind of investigation, however, involves the engine to a secondary degree, in fact for such work the engine is often replaced by an electric motor or other power source. These tests are mainly concerned with the development of transmission systems and are thus rather outside the scope of this book. The basic principles of oscillation in engine–dynamometer systems are dealt with in Chapters 5 and 8. See also ref. 1.

Time scale 0.5–2.0 sec

This time scale is characteristic of gear shifts and correct behaviour in this range is critical in a system called upon to simulate these events. This is a particularly demanding area, as the profile of gearshifts is extremely variable. At one extreme we have the fast automatic changes of a race car gearbox, at the other gear changes in the older commercial vehicle, which may take more than 2 seconds, and in between the whole range of individual driver characteristics, from aggressive to timid.

Whatever the required profile, the dynamometer must impose zero torque on the engine during the period of clutch disengagement and this calls for precise following of the 'free' engine speed. The accelerating/decelerating torque required is proportional to dynamometer inertia and the rate of change of speed:

$$T = I \, d\omega/dt$$

where T = torque required, Nm, I = dynamometer inertia, $kg\,m^2$, and $d\omega/dt$ = rate of speed change, rad/s. This requirement is a major factor in the choice of dynamometers for transient testing, see p.272.

A gear change involves throttle closing, an engine speed change, up or down, of perhaps 2000 rev/min, and the re-application of power. Ideally, at every instant during this process the rate of supply of fuel and air to the engine cylinders and the injection and ignition timings should be identical with those corresponding to the optimized steady state values for the engine load and speed at that instant. It is possible to calculate the corresponding 'ideal' emission quantity for the gear change, see for example ref. 2, and to compare this with the performance actually achieved. It will be clear that this optimizing process makes immense demands on the engine management system (also on the emissions instrumentation, see p.255).

A special problem of engine control in this area concerns the response of a turbocharged engine to a sudden demand for more power. An analogous problem has been familiar for many years to process control engineers dealing with the management of boilers. If there is an increase in demand the control system increases the air supply before increasing the fuel input; on a fall in demand the fuel input is reduced before the air. The purpose of this 'air first-fuel first' system is to ensure that there is never a deficiency of air for proper combustion.

In the case of a turbocharged engine only specialized means, such as variable turbine geometry, can increase the air flow in advance of the fuel flow. A reduction in value from the (optimized) steady state value, with consequently increased emissions, is an inevitable concomitant to an increase in demand. The control of maximum fuel supply during acceleration is a difficult compromise between performance and driveability on the one hand and a clean exhaust on the other. Figure 15.1, from ref. 3, is a vivid illustration of the control problems associated with the turbocharged engine when driving through a conventional gearbox. This shows the variations in engine torque and turbocharger speed during the acceleration of a 30 tonne truck driven by a 10 litre diesel engine.

A further important aspect of vehicle performance that falls within this time scale concerns the onset of skidding and wheel spin and the corresponding reactions of the engine and braking control systems.

Time scale 1.0–10.0 sec

Here we are concerned with driver-initiated changes other than gear shifts, such as throttle opening and brake application. The same considerations discussed above apply, but to a less acute degree.

Time scale 1–10 min

The warm–up time of an engine lies within this time scale. The study of emissions and of the performance of engine cooling systems, lubricating oil systems, exhaust catalysers, etc. during warm-up is a major development concern, but does not present particular problems of instrumentation.

The road load equation

An engine driving an industrial machine such as a generator or a pump usually runs at constant speed and its control system is required only to deal with comparatively slow changes in torque demand. At a higher level of complexity a marine engine runs at varying speeds but torque and speed are linked by a simple relationship: the 'propeller law'. In the case of a vehicle engine the situation is vastly more complicated. The power demanded from the engine depends on a large number of factors, which depend on the driving conditions and are unique to any particular vehicle.

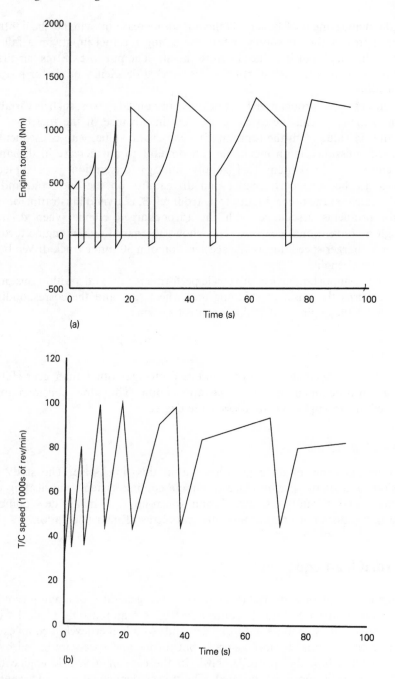

Figure 15.1. *Turbocharged engine: performance during gear changes: (a) engine torque and (b) turbocharger speed versus time.*

The behaviour of a vehicle under road conditions is described by the *road load equation*. This defines the traction or braking force that is called for under all possible conditions, which include:

- steady travel at constant speed on a level road
- hill climbing and descent
- acceleration, over-run and braking
- transitions between any of the above
- effects of wind, load, towed load, tyre pressure, etc.

The road load equation for a given vehicle defines the *tractive force/retarding force F* Newton that must be applied in order to achieve a specified response to these conditions. It is a function of the following parameters:

vehicle specific:	mass of vehicle	M	kg
components of resistance: rolling		a_0	N
	speed dependent	$a_1 V$	N
	aerodynamic	$a_2 V^2$	N
external:	speed	V	m/s
	road slope	θ	rad

The usual form of the road load equation is:

$$F = a_0 + a_1 V + a_2 V^2 + M \, dV/dt + Mg \sin \theta \qquad (15.1)$$

where $M dV/dt$ = force to accelerate/brake vehicle
$Mg \sin \theta$ = hill climbing force

More elaborate versions of the equation may take into account such factors as wheel spin and cornering.

The practical importance of the road load equation lies in its application to the simulation of vehicle performance: it forms the link between performance on the road and performance in the test department.

To give a 'feel' for the magnitudes involved, the following equation relates to a typical four-door saloon of moderate performance, laden weight 1600 kg:

$$F = 150 + 3V + 0.43 \, V^2 + 1600 \, dV/dt + 1600 \times 9.81 \sin \theta$$

The various components that make up the vehicle drag are plotted in Figure 15.2. Fig. 15.2(a) shows the level road performance and makes clear the preponderant influence of wind resistance. The road load equation predicts a power demand at the road surface of 14.5 kW at 60 m.p.h., rising to 68.8 kW at 50 m/s (112 m.p.h.) for a level road. These demands are dwarfed by the demands made by hill climbing and acceleration. Thus, Fig. 15.2(b), to climb a 15 % slope at 60 m.p.h. calls for a total power input of 14.5 + 63.2 = 77.7 kW while, Fig. 15.2(c), to accelerate from 0–60 m.p.h. in 10 seconds at constant acceleration calls for a maximum power input of 14.5 + 115 = 129.5 kW.

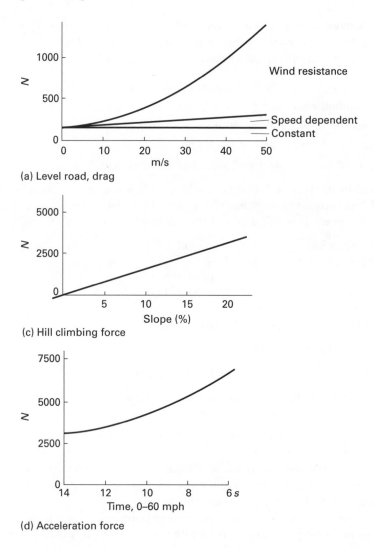

(a) Level road, drag

(c) Hill climbing force

(d) Acceleration force

Figure 15.2.   *Vehicle drag: four-door saloon, laden weight 1600 kg (a) Level road performance; (b) Hill climbing; (c) Acceleration force.*

Dynamometers for transient testing: special requirements

It will be apparent that to replace a vehicle drive line, which includes a clutch, by a dynamometer that is permanently coupled to the engine, and then to expect that the pattern of loading imposed by the dynamometer will exactly mimic the pattern imposed by the vehicle makes very severe demands on the dynamometer and its control system.

While until recently d.c. dynamometers and drive systems have largely monopolized dynamic testing applications, recent developments in a.c. machines and drive systems have led to their increasing use in this field. d.c. machines require relatively simple drive systems, but are more expensive to construct, require regular commutator maintenance, and, of particular importance, have higher moment of inertia than equivalent a.c. machines.

a.c. dynamometers are themselves simple in construction and cheap to maintain and may be run at higher speeds than d.c. machines of comparable power, but the associated vector drive systems are complex and expensive. Typically, a d.c. machine may have four times the moment of inertia of an equivalent a.c. machine, with consequent more limited maximum rates of speed change.

In the vehicle, gear changes are characterized by the instantaneous disappearance of load on the engine, followed milliseconds later by the reimposition of a different load at a different speed. A dynamometer required to simulate this process must be capable of exceedingly rapid acceleration and deceleration and rates of speed change of up to 18 000 r.p.m./s are not unknown.

The evolution of computerized control systems, mainly a.c., capable of simulating in real time the load pattern experienced by the engine in the vehicle, continues. However, interactions between this system, the test cell computer and the engine management unit can be extremely complex and may give rise to undesirable and unexpected results.

Safety may be jeopardized, for example by the dynamometer motoring a failing engine in order to maintain the speed demanded by the programme and there can be a real danger, with the interactions between control systems measured in nanoseconds, of one or more systems becoming 'confused', a danger that did not arise when most testing involved high inertia d.c. dynamometers and analogue control systems.

Summary

Transient tests, which are almost entirely concerned with vehicle engines, are classified in terms of the time scale involved, and range from the study of high-frequency transients such as transmission judder to comparatively slow changes associated with engine warm-up. Engine calibration, homologation and bench marking are defined, and the road load equation is derived and illustrated by an example. The special features of dynamometers for transient testing are described.

References

1. *Symposium (on) Dynamic Testing* (1990) A.V.L. List GmbH, Graz, Austria.
2. Beaumont, A.J. *et al.* (1990) *Signal Reconstruction Techniques for Improved Measurement of Transient Emissions*, S.A.E. Technical Paper, No. 900233.
3. Wallace, F.J. and Desquan, Z. (1990) On-road performance of two alternative engine transmission systems for heavy vehicles. *Proc. I. Mech. E.* C405/038.
4. Von Thun, H.A. *et al.* (1988) *A New Dynamometer Test Rig to Develop Drive Lines for All-wheel Driven Vehicles*, S.A.E. Technical Paper, No. 881736.

Further reading

Pilley, A.D. *et al.* (1989) *Optimisation of Heavy-duty Diesel Engine Transient Emissions by Advanced Control of a Variable Geometry Turbocharger*, S.A.E. SP-781 Electronic, Diesel Engine Controls, Paper No. 890395.
Martyr, A.J. *et al.* (1990) *The Requirements and Development of Dynamometer and Control Equipment for Testing Engines in the Transient Phase*, I.Mech.E. Seminar, Engine Transient Performance, 7 November 1990.

16 Chassis or rolling road dynamometers

This Chapter deals with an area of testing activity which has increased enormously in importance during the last quarter of the 20th Century, and that shows every sign of continuing to expand. It is exclusively concerned with road vehicles and their engines and its expansion has been driven by a number of developments:

Technical: the need to perfect the complex engine and whole vehicle management systems that are demanded by the modern road user, the economical performance of endurance testing, investigations of noise, vibration and harshness, emissions testing, checking performance under extreme conditions, etc.

Legislative: the need to comply with ever more exacting statutory requirements governing such factors as vehicle safety, brake performance, NVH, EMC, fuel consumption, exhaust emissions, etc. (see Homologation, p. 266).

These requirements have called into existence a hierarchy of dynamometers and test chambers of increasing complexity:

- brake testers
- in-service tuning and assessment
- end-of-line production testing
- emissions testing installations
- mileage accumulation facilities
- anechoic and NVH and EMC test chambers
- climatic chambers and wind tunnels
- independent wheel dynamometers.

In this Chapter these various types are considered, with some discussion of the design problems involved.

The basic theory behind much of this work, the *road load equation*, is dealt with in Chapter 15.

Genesis of the rolling road dynamometer

The idea of running a complete vehicle under power while it was at rest was first conceived by railway locomotive engineers before being adopted by the road vehicle industry. The last steam locomotives built in the UK were tested on multiple roll units, with large eddy-current dynamometers connected to each driven roll, the tractive force being measured by a mechanical linkage and spring balance. To an observer close to the locomotive it must have presented an awe-inspiring sight, running at full power and a wheel speed of 90 mph.

Today the chassis dynamometer is used almost exclusively for road vehicles, although there are special machines designed for fork-lift and articulated off-road vehicles. The advantages to the designer and test engineer of having such facilities available are obvious; essentially they allow the static observation of the performance of the complete vehicle while it is, in most respects, effectively in motion.

Before about 1970 most machines were comparatively primitive 'rolling roads', characterized by having rollers of rather small diameter, which thus poorly simulated the tyre contact conditions and rolling resistance experienced by the vehicle on the road, and with various fairly crude arrangements for applying and measuring the torque resistance. A single fixed flywheel was commonly coupled to the roller to give a simulation of an average vehicle inertia.

The main impetus for development came with the rapid evolution of emissions testing in the seventies. The diameter of the rollers was increased, to give more realistic traction conditions, while trunnion mounted d.c. dynamometers with torque measurement by strain gauge load cells and more sophisticated control systems permitted more accurate simulation of road-load sequences. The machines were provided with a range of flywheels to give steps in the inertia. Precise simulation of the vehicle mass was achieved by 'trimming' the effective inertia electronically. In recent years there has been a great increase in activity in the area of vehicle homologation, p.266, and this has called for a large increase in the number of machines in service. These machines, based on a simple single roll of large diameter, are true chassis dynamometers capable of putting the vehicle through the complex driving sequences demanded by legislation, p.249.

Brake testers

These simple machines, commonly installed in service garages, are used mainly for the statutory 'in service' testing of cars and commercial vehicles. They are installed in a shallow pit, and consist essentially of two pairs of rollers, typically of 170 mm diameter and having a coated surface to give a high coefficient of adhesion with the vehicle tyre. The rollers are driven by geared variable speed motors and two types of test are performed. Either the vehicle brakes are fully applied and torque increased until they slip, or the rollers are driven at a low surface speed, typically 2 km/h, and the relation between brake pedal effort and braking force is measured.

The testing of a modern ABS brake system requires a separately controlled roller for each wheel and a much more complicated interaction with the vehicle braking system. This calls for computerized control of the tester and in many vehicle management systems a plug is provided for connection of the brake system to a suitable brake tester computer.

Rolling roads for in-service tuning and assessment

For the typical vehicle maintenance garage the installation of a rolling road represents a considerable investment, only justified if the garage has a good market for its expertise in vehicle tuning or other specialized skills. To cater for this potentially very large market a number of manufacturers produce complete installations based on small diameter rollers and thus requiring minimum sub-floor excavation. This market tends to be dominated by American suppliers, and will grow rapidly, since the US Environmental Protection Agency (EPA) now calls for annual emissions testing of all vehicles, the test to be based on a chassis dynamometer cycle (IM 240).

Road load simulation capability at this level is usually limited to that possible with a choice of one or two flywheels and a comparatively simple control system; most of the testing is concerned with optimizing (tuning) the engine management system and straightforward maximum power certification. The tests have to be of short duration to avoid overheating of engine and tyres.

In-service rolling roads for testing large trucks are confined almost exclusively to the USA. They are based on a single large roller capable of running both single and double axle tractor units. The power is usually absorbed by a portable water brake, such as that shown in Fig. 2.1, which may also be used for direct testing of engines, a useful facility in a large agency test facility.

Rolling roads for end-of-line production testing

Test rigs for this purpose range from simple roller sets to multi-axle units with inter-connected roll sets and road load simulation. Certain special design features are required and any specification for such an installation should take them into account:

- the design and construction must minimize the possibility of damage from vehicle parts falling into the mechanism. It is quite common for fixings to shake down into the rolling road, where they can cause damage if there are narrow clearances or converging gaps between, for example, the rolls and lift-out beam.
- the vehicle must be able to enter and leave the rig quickly yet be safely restrained during the test. The usual configuration is for all driven wheels to

run between two rollers, Fig. 16.1. Between each pair of rolls there is a lift-out beam that allows the vehicle to enter and leave the rig. When the beam descends it lowers the wheels between the rolls and at the same time small 'anti - climb out' rollers swing up fore and aft of each wheel. At the end of the test, with the wheels at rest, the beam rises, the restraining rollers swing down, and the vehicle may be driven from the rig.

- to restrain the vehicle from slewing from side to side to a dangerous extent, specially shaped side rollers are positioned between the rolls at the extreme width of the machine where they will make contact with the vehicle tyre to prevent further movement.

- particular care should be taken with the operating procedures if 'rumble strips' are to be used. These are raised strips or keys sometimes fitted to the rolls in line with the tyre tracks, their purpose being to excite vibration to assist in the location of vehicle rattles. However, if the frequency of contact with the tyre resonates with the natural frequency of the vehicle suspension this can give rise to bouncing of the vehicle and even to its ejection from the rig. They can also give rise to severe shock loads in the roll drive mechanism.

Rigs designed to absorb power from more than one axle may have to be capable of adjustment of the centre distance between axles, which increases the complexity of the necessary shaft or belt connection between the roller sets. Special purpose roller rigs for checking wheel alignment sometimes form part of the end-of-line test equipment.

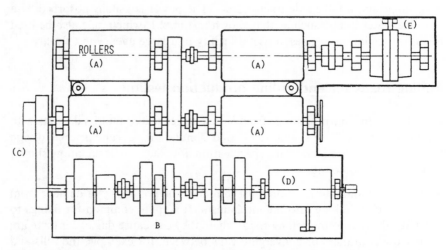

Figure 16.1. *Schematic plan view of four roller type chassis dynamometer. (A) Rollers, (B) flywheel set, (C) clutchable chain drive, (D) d.c. machine, (E) eddy current brake.*

Chassis dynamometers for emissions testing

The standard emissions tests developed in the U.S.A. in the seventies were based on a rolling road dynamometer developed by the Clayton Company and having twin rolls of 8.625 in (220 mm) diameter. This machine became a *de facto* standard despite its limitations, the most serious of which was the small roll diameter, which resulted in tyre contact conditions much different from those on the road.

Later models use rollers of 500 mm dia. coupled to a four-quadrant d.c. dynamometer with de-clutchable flywheels to simulate vehicle inertia, Fig. 16.2. The machine is designed for roller surface speeds of up to 160 km/h and the digital control system is capable of precise road load control, including gradient simulation. Such machines must be manufactured to a very high standard to keep vibration to an acceptable level. It is possible to abuse them, notably by violent brake application which can induce very high stresses in the roller drive mechanism.

More recent recommendations from the US Environmental Protection

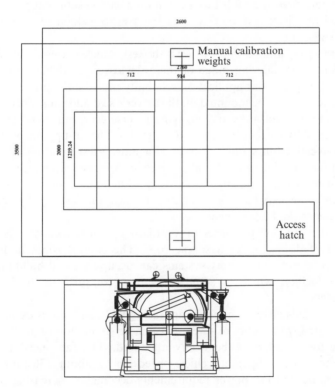

Figure 16.2. *EPA approved emissions dynamometer with 48 in. rolls, centre motor and vehicle centring roller linkage*

Agency have approved machines having a single pair of rollers of 48 in dia. and 100% electronic inertia simulation. The majority of chassis dynamometers manufactured world-wide in the last 25 years have been used for emissions testing and homologation.

Mileage accumulation facilities

As a direct result of emissions regulations and the practice of homologation it has become necessary for manufacturers to check on the change and probable deterioration in emissions performance after the vehicle has been driven for up to 80 000 km. The cost and the physical strain of using human drivers on test tracks or public roads are high so special chassis dynamometers have been developed for running the specified sequences under automatic control, commonly for a period extending to 12 weeks.

Mileage accumulation dynamometers comprise a single large roll, 1.0 to 1.2 m diameter, directly coupled to a d.c. motor having sufficient power to run the required repetitive test sequence; for the average saloon car a capacity of 150 kW is sufficient.

In order to fully automate the process the test vehicle is usually fitted with a robot driver. A bewildering range of these devices is available on the world market: for this particular application reliability must be the prime consideration. Some devices can be mounted on the driver's seat and some are capable of also operating the footbrake and engaging reverse gear, but these refinements are not required by current legislative drive cycles. The set-up time of the robot may be appreciable but should be assessed against a test duration of 12 weeks or more. If wear or deterioration takes place in the robot mechanism or in the vehicle controls, for example by change in the clutch 'bite' position, the control system should be capable of automatic rectification, usually by means of periodic 'relearn' and adjustment cycles.

Due to the obvious problems of ventilating such a facility it is usually located out of doors, under a simple roof. The control room housing the computerized control and safety systems can occupy a small building at one end of the facility, which in some cases may comprise up to 10 chassis dynamometers.

As the vehicle is at rest a cooling fan, facing the front end, is essential. The fan will be fitted with a duct to give a reasonable simulation of air flow under road conditions, and must be firmly anchored. The fan speed is usually controlled to match apparent vehicle speed up to about 130 km/h; above that speed noise and fan power requirements become an increasing problem. Since mileage accumulation facilities usually run 24 hours a day they must be suitably shielded if residential areas may be affected.

Special purpose chassis dynamometers

Noise, vibration and harshness (NVH), electro-magnetic compatibility (EMC) rigs and anechoic chamber dynamometers

Some features of anechoic cell design are discussed in Chapter 5, p.87. A critical requirement is that the chassis dynamometer should itself create the minimum possible noise. The usual specification calls for the noise level to be measured by a microphone located 1 m above and 1 m from the centreline of the rolls. Typically, the required sound level when the rolls are rotating at a surface speed of 100 km/h will be in the region of 50 dBA.

To reduce the contribution from the dynamometer motor and its drive system it is usually located outside the main chamber in its own sound-proofed compartment and connected to the rolls by way of a long shaft (the design of these shafts can present problems and lightweight tubular carbon fibre shafts are sometimes used). The dynamometer motor will inevitably require forced ventilation and the ducting will require suitable location and treatment to avoid any contribution of noise.

The use of hydro-dynamic bearings on quiet roll systems has declined as their expected advantages in terms of reduced roll noise have not been realized in practice while the noise from the necessary pressurized lubrication system and the added complexity are disadvantages.

For a well designed chassis dynamometer the major source of noise will be the windage generated by the moving roll surface and this is not easily dealt with. Smooth surfaces and careful shrouding can reduce the noise generated by the roller end faces, but there is inevitably an inherently noisy jet of air generated where the roller surface emerges into the test chamber. If the roller has a roughened surface to simulate road conditions, p.287, the problem is exacerbated. The noise spectrum generated by the emerging roller is influenced by the width of the gap and in some cases adjustment is provided at this point.

The flooring over the dynamometer pit, usually of steel plate, must be carefully designed and appropriately damped to avoid resonant vibrations. A particular feature of NVH test cells is that the operators require access to the underside of the vehicle so that adjustments to the instrumentation and the vehicle can be made. This is usually by way of a trench at least 1.8 m deep, lying between the vehicle wheels. For some NVH work the chassis dynamometer is not used and the vehicle wheels preferably rest on a concrete surface.

This access may be difficult to achieve in a facility intended also to accommodate a chassis dynamometer and it is recommended that whenever new facilities are being planned a careful survey be made of the requirements of the future users.

Testing for electro-magnetic compatibility (EMC)

European Safety Standards, discussed briefly in Chapter 3, p.42, include regulations governing the acceptable levels of electro-magnetic interference produced by road vehicles and, conversely, the vulnerability of vehicle electronic systems to powerful directional beams of electromagnetic energy, such as may be produced by civil and military aviation systems. To deal with this problem a few units have been designed with the chassis dynamometer mounted on a large turntable capable of being rotated through 360°.

Climatic test cells for vehicles

There is an increasing requirement for this kind of facility for development work associated with various aspects of performance under extreme climatic conditions. Subjects include driveability, cold starting, fuel waxing and vapour lock, air conditioning and the ubiquitous topic of emissions.

The design of air conditioning plants for combustion air has been discussed in Chapter 6, p.110; the design of an air conditioned chamber for the testing of complete vehicles is a very much larger and more specialized problem. Figure 16.3 is a schematic drawing of such a chamber, built in the form of a recirculating wind tunnel so that the vehicle on test, mounted on a chassis dynamometer, can be subjected to oncoming air flows at realistic velocities, and in this particular case over a temperature range from $+40°C$ to $-30°C$.

The outstanding feature of such an installation is its very large *thermal mass* and, since it is usually necessary, from the nature of the tests to be performed, to vary the temperature of the vehicle and the air circulating in the tunnel over a wide range during a test, this thermal mass is of prime significance in sizing the associated heating and cooling systems.

The thermodynamic system of the chamber is indicated in Fig.16.3 and the energy inflows and outflows during a particular cold test, in accordance with the methods described in Chapter 1, were as follows:

The thermal capacity (thermal mass) of the various elements of the system is defined in terms of the energy input required to raise the temperature of the element by $1°C$. The various terms are derived as follows:

In		*Out*	
H_a make-up air	20 kW	H_r refrigerator circuit	495 kW
H_v vehicle convection/radiation	100 kW		
H_f tunnel fan power	154 kW		
H_w walls, etc., by difference	221 kW		
	495 kW		495 kW

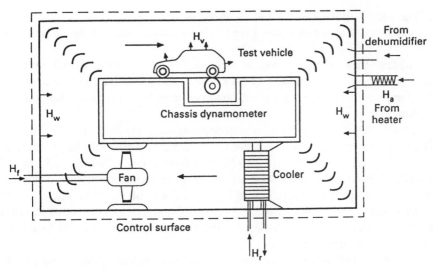

Figure 16.3. *Climatic test cell*

C_a air content of chamber, return duct, etc. volume approximately $550 \, \text{m}^3$ approximate density $1.2 \, \text{kg/m}^3$ $C_p = 1.01 \, \text{kJ/kg.K}$ thermal mass $= 550 \times 1.2 \times 1.01 = 670 \, \text{kJ/deg C}$

C_v vehicle, assume a commercial vehicle, weight 3 tonnes specific heat of steel approximately $0.45 \, \text{kJ/kg.K}$ thermal mass, say $3000 \times 0.45 = 1350 \, \text{kJ/deg C}$

C_e fan, cooler matrix, internal framing etc. estimated at $2480 \, \text{kJ/deg C}$

C_s concrete structure of chamber. This was determined by running a test with no vehicle in the chamber and a low fan speed. The rate of heat extraction by the refrigerant circuit was measured and a cooling curve plotted. Concrete is a poor conductor of heat and the coefficient of heat transfer from surfaces to air is low. Hence during the test a temperature difference between surfaces and wall built up but eventually stabilized; at this point the rate of cooling gave a true indication of effective wall heat capacity. The test showed that the equivalent thermal mass of the chamber was about $26\,000 \, \text{kJ/deg C}$, much larger than the other elements. Concrete has a specific heat of about $8000 \, \text{kJ/m}^3 \, \text{deg C}$, indicating that the 'effective' volume involved was about $26\,000/8000 = 3.2 \, \text{m}^3$

The total surface area of the chamber is about $300\,m^2$, suggesting that a surface layer of concrete about 1 cm thick effectively followed the air temperature.

Adding these elements together we arrive at a total thermal mass of $30\,500\,kJ/deg$ C. Then since $1\,kJ = 1\,kW\text{-sec}$, we can derive the rate at which the temperature of the air in the chamber may be expected to fall for the present case.

This shows that the 'surplus' cooling capacity available for cooling the chamber and its contents, H_w, amounts to $221\,kW$. This could be expected to achieve a rate of cooling of $221/30\,500 = 0.0072$ deg C/sec, or $1°C$ in 2.3 minutes. However this is the final rate when the temperature difference between walls and air had built up to the steady state value of about $20°C$. The observed initial cooling rate is much faster, about $1.5°C/min$. Heating presents less of a problem, since the heat released by the test vehicle and the fan assist the process rather than opposing it.

A further effect is associated with the moisture content of the tunnel air. On a warm summer's day this could amount to about 10 kg of water and during the cooling process this moisture would be deposited, mostly on the cooler fins, where it would eventually freeze, blocking the passages and reducing heat transfer. To deal with this problem it is necessary to include a dehumidifier to supply dried air to the tunnel circuit.

Chassis dynamometers intended to operate over the temperature range $+40°C$ to $-10°C$ call for no special features, apart from sensible precautions to deal with condensation. For temperatures below about $-25°C$ certain special features will be required. The dynamometer motor should be isolated from the cold chamber and operate at normal temperatures while the cold components, rolls and shaft, must be of steel having adequate low-temperature strength, to avoid the risk of brittle fracture. Shaft bearings and parking brake systems may require trace heating.

Independent wheel dynamometers

A limitation of the conventional rolling road dynamometer is that it is unable to simulate cornering, during which the wheels rotate at differing speeds, wheel spin and skidding. With the increased adoption of electronically controlled traction and braking (ASR and ABS) there is a growing requirement for test beds that can simulate these conditions. There are two types of solution.

Four roll set rolling roads

These machines range from 'end of line' rigs consisting of four sets of independent double-roller units to complex development test rigs having steered articulation of each roll unit. The production rigs permit the checking of onboard vehicle control systems, transducers and system wiring by simulation of differential resistance and speed of rotation.

Wheel substitution dynamometers

A major problem faced by engineers carrying out NVH testing on a chassis dynamometer is that tyre noise can dominate sound measurements. One answer is to absorb the power of each drive wheel with an individual dynamometer. If the wheel hub is modified the vehicle can still sit on its tyres, giving approximately the correct damping effect, while the tyre contact and windage noise is eliminated. Alternatively the wheels may be removed and the individual wheel dynamometers used to support the vehicle. The dynamometers should be four-quadrant machines, to simulate both driving and coasting/braking conditions. Both a.c. and d.c. machines have been used for this application, although most recent designs have been of the hydrostatic type.

Special features of rolling road test cells

The engineer called upon to draw up a specification for a chassis dynamometer test facility should be aware of these requirements.

Pit design

The sub-floor pit in which a chassis dynamometer is to be installed must be built to a standard of accuracy rather higher than is usual in civil engineering construction, since there are a number of interfaces with mechanical and structural components. The steel floor plates will be lifted from time to time and it is recommended that the flooring contractor should supply the pit edging, to be cast into the pit walls under his supervision. The pit should be built with a slight excess depth so that the dynamometer may be correctly aligned with the floor, by machine mounts or shim plates.

The pit requires ventilation to avoid possible build up of inflammable vapour, and the system should be interlocked with the cell safety system as recommended for engine test cells, p.21. In addition there must be a well-thought-out 'lock out' system to ensure that all personnel are outside restricted areas before start-up, particularly important in large facilities where the operator may not be able to see all areas.

Precautions must also be taken to avoid the possibility of flooding, which could cause expensive damage. The pit floor should be laid to a fall so that spillage runs to a drain point, where a float-controlled scavenge pump may be necessary, depending on the general drainage system layout.

Tyre burst detectors

On all mileage accumulation rigs and others undertaking prolonged automatic test sequences there should be some form of detector that can safely shut down the whole system in the event of a tyre deflating. The most common form is a limit switch mounted on a floor stand with a long probe running under the car at its mid-point with limit switches fitted one on each side. Any deflation will cause the vehicle body to drop, shutting down the test.

Cable layout

The general treatment of this subject in Chapter 3 applies in the present case, but special attention should be given to the avoidance of electrical interference from power cables with instrumentation lines.

Loading and emergency brakes

It must be possible to lock the rolls to permit loading and unloading of the vehicle and it is usual to fit disc brakes to the roll shafts, of sufficient power to resist the torques associated with driving the vehicle on and off the rig; these brakes, either pneumatic or hydraulic, are controlled by the operational and safety instrumentation and are programmed to be normally 'on', except when the system is activated and ready to run. It is not considered good practice to rely on these brakes to bring the rig to rest in the case of an emergency: they should only be used to supplement the braking effort of the main drive system.

Vehicle restraints

The test vehicle must be adequately restrained against fore and aft motion under the tractive forces generated, and against slewing or sideways movements caused by the wheels being set at an angle to the rig axis. Vehicle restraint is less of a problem when the powered wheels are located between a pair of rolls rather than resting on a single roller; however paired rolls are only used for short duration test work because of their limited diameter and hence poor contact pattern. They are generally used for end-of-line testing, where indeed they are usually relied on as the only means of restraint, in view of the need for rapid loading and unloading.

   The cell floor must be provided with strong anchorage points. There are many different designs of vehicle restraint equipment, ranging from the use of large pneumatic bags to simple chain attachments. Three different types of restraint system may be distinguished:

- a vehicle with rear wheel drive is the easiest type to restrain. The front wheels can be prevented from moving by fore and aft chocks linked by a tie bar. The rear end may be prevented from slewing by two high strength luggage straps with integral tensioning devices; these should be arranged in a cross-over configuration with the floor fixing points outboard and to the rear of the vehicle.

- front wheel drive vehicles need careful restraint since any movement of the steering mechanism with the rolls running at speed will lead to violent yawing. Restraints may be similar to those described for rear wheel drives but with the straps at the front and chocks at the rear. The driver in manned runs should be familiar with the handling characteristics of the vehicle in normal operation while for unmanned operation the steering wheel should be locked, otherwise disturbances such as a burst tyre could have serious results. Tyre pressures, usually set above the normal level, should be carefully equalized.
- for four wheel drive vehicles it is usual to rely on cross-over strapping at both ends. It is good practice to tie the rear end first and then to drive the vehicle slowly ahead with the steering wheel loosely held. In this way the vehicle should find its natural position on the rolls and can be restrained in this position, giving the minimum of tyre scrubbing and heating.

Roll surface treatment

In the past most twin roll and many single roll machines had no special treatment applied to the roll surfaces, which had a normal finish machined surface. Brake testers and some production rigs have a high-friction surface, which can give rise to severe tyre damage if the machine is used incorrectly. The roll surface of modern single roll machines is commonly sprayed with a fine grained tungsten carbide coating.

'Road shells'

For development work, and particularly for NVH development, it is sometimes desirable to have a simulated road surface attached to the rolls. Usually the required surface is a simulation of quite coarse stoned asphalt; this generates the typical high frequency 'white noise' when a vehicle is running on it. However a roughened roll surface greatly increases the windage noise of the rig itself.

Road shell design and manufacture is difficult, while it is usually necessary to limit roll speed when shells are in use. The shells should ideally be made up in four or more segments of differing lengths and with junction gaps that are as small as possible and cut helically. It is rarely required to induce low frequency vibrations in the test vehicle; if it is required it should be generated by randomly distributed variations in roll surface height to avoid possible resonances in the vehicle suspension.

The most usual techniques for producing road surfaces for attachment to chassis dynamometer rolls are:

- Detachable fibreglass road shells having an accurate moulding of a true road surface. These shells are usually thicker than the aluminium type, below, and in both cases the floor plates must be adjustable in order to accommodate them.
- Detachable cast aluminium alloy road shells, made in segments that may be bolted to the outside diameter of the rolls. The surface usually consists of parallel sided pits that give an approximation to a road surface.
- A permanently fixed road surface made up of actual stones bonded into a rubber belt which is itself bonded to the roll surface.

Most road shells are not capable of running at maximum rig speed because thay are difficult to fix and to balance, also they may not be capable of transmitting full acceleration torque. It is therefore necessary to provide safety interlocks so that the computerized speed and torque alarms are set to appropriate lower levels when shells are in use. Road shells also change the effective rolling radius and the base inertia of the dynamometer, requiring appropriate changes in the software constants.

Driver aids

The control room needs to be in touch with the driver in a test vehicle on a chassis dynamometer, either by voice transmission or visually by way of an aid screen which must be easily visible to him.

A major function of such a screen will be to give the driver instructions for carrying out standardized test sequences, for example production test programmes or emissions test sequences. This kind of VDU display is often graphical in form, showing for example the speed demanded and the actual speed achieved. Since in the case of emissions tests the test profile must follow that prescribed within defined limits it is usual to include an error checking routine in the software to avoid wasted test time. The screen may also be used to transmit instructions to supplement information conveyed by the two-way voice link.

Fire suppression

The risk of a vehicle fire during chassis dynamometer running is quite high, since the air cooling, even with supplementary fans, is likely to be less than that experienced on the road. Underfloor exhaust systems in particular can become very hot and could ignite fuel vapour, whatever its source. Fire fighting is more difficult than in an ordinary test cell both because of the large size of rolling road cells and the more difficult access to the seat of the fire, which may be within the vehicle body.

All vehicle test facilities should be equipped with substantial hand held or hand operated fire extinguishers and staff should be trained in their use. An alternative is to fit the test vehicle with a system of the type designed for rally cars which enables the driver or control room to flood the engine compartment with foam extinguishant. Automatic systems of the type used in engine test cells, p.21, are less effective in vehicle cells in view of the greater difficulty in ensuring that all personnel, including the driver, have been evacuated before they are activated. The modern trend is towards water fog suppression, which may include discharge nozzles mounted beneath the vehicle and thus near the seat of most potential fires.

There must be a clear and unimpeded escape route and the impairment in vision from possible steam or smoke must be taken into account, particularly in the case of anechoic cells, where the door positions should not be camouflaged by the coned surface.

The effect of roll diameter on tyre contact conditions

The subject of tyre rolling resistance has given rise to an extensive literature, which is well summarized in ref. 1. The bulk of the rolling resistance is the consequence of hysteresis losses in the material of the tyre and this gives rise directly to heating of the tyre. A widely accepted formulation describing the effect of the relative radii of tyre and roll is:

$$F_{xr} = F_x(1 + r/R)^{1/2} \tag{16.1}$$

where:

F_{xr} = rolling resistance against drum
F_x = rolling resistance on flat road
r = radius of tyre
R = radius of drum

Figure 16.4(a) shows the situation diagrammatically and Fig. 16.4(b) the corresponding relation between the rolling resistances: this shows that rolling resistance (and hence hysteresis loss) increases linearly with the ratio r/R. For tyre and roller of equal diameter the rolling resistance is $1.414 \times$ resistance on a flat road, while for a tyre three times the roller diameter, easily possible on a brake tester, the resistance is doubled. A typical value for the 'coefficient of friction' (rolling resistance/load) would be 1% for a flat road.

To indicate magnitudes involved, a tyre bearing a load of 300 kg running at 17 m/s (40 mph) could be expected to experience a heating load of about 500 W on a flat road, increased to 1 kW when running on a roller of one third its diameter.

Clearly the heating effects associated with small diameter rolls are by no means negligible.

Tyres used even for a short time on rolling roads may be damaged by these

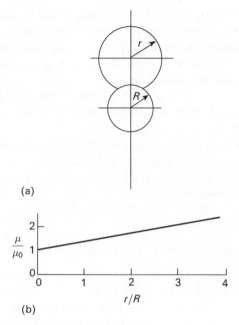

(a)

(b)

Figure 16.4. *Vehicle restraint device*

heating effects, which are aggravated by the fact that the roll will also be heated in the course of the run. All tyres used for any but the shortest duration must be specially marked and should be changed before the vehicle is allowed on public roads.

Specification, accuracy and calibration

A typical specification for a chassis dynamometer will include the following information as a minimum:

- base inertia[*]
- minimum simulated inertia
- maximum simulated inertia
- continuous rated power, absorbing mode
- continuous rated power, motoring mode
- torque range, absorbing and motoring
- number of rolls

[*] It is common practice to use the term 'inertia' to mean the 'simulated', or equivalent vehicle inertia, rather than the moment of inertia of the rotating oomponents. The inertia is thus quoted in kg rather than in $kg\,m^2$.

- roll diameter with tolerance (typically $< \pm 0.005$ in. on a 48 in. roll)
- roll concentricity (typically $< \pm 0.005$ in. on a 48 in. roll)
- maximum speed, drum rev/min and roll surface speed
- maximum permitted acceleration
- maximum time to stop from maximum rated speed under regenerative braking
- type of drive protection system

These parameters determine the basic geometry of the rig and the size of motor and its associated variable speed drive system.

Where a purpose of the dynamometer is to carry out homologation tests these should be clearly specified, together with the range of vehicles to be covered; the acceptance procedure should include performance of these tests.

The essential measurements performed by a chassis dynamometer, equivalent to the measurement of torque and speed in a conventional machine, are roller surface speed and the force (the tractive effort at the roll periphery) transmitted to or from the rolls.

This force is usually not measured directly, but derived from a torque measurement at the motor, performed either in the conventional manner by utilizing a trunnion mounted motor with torque arm and load cell, or by some form of torque shaft dynamometer. Whatever method is used, there will be an in-built error arising from bearing losses and roll windage: these losses can be measured by running the machine through the speed range with no vehicle present and noting the necessary driving torque, which should be added to the observed torque when absorbing power and subtracted when braking, to give the true tractive force at the roll rim. (This is not the complete answer as there will be a further, unquantified, loss due to roll bearing friction when the rolls are under load).

Some dynamometers include these corrections in the stated dynamometer constants. See p.161. for a detailed discussion of torque calibration procedures. This is often not a simple matter in the case of chassis dynamometers as the torques involved can be very substantial and space very restricted: the preferred method using graduated weights to apply torque in ten steps may not be practicable and dynamometers are often supplied with a single mass corresponding to full torque plus a few small weights adding up to about 10% of full scale torque.

The accuracy of calibration demanded and the calibration methods to be used may be included in emissions legislation relevant to chassis dynamometer testing as part of the general tightening of standards.

Equivalent road speed measurement will usually be by means of a high resolution optical encoder on the roll shaft, thus introducing a possible source of error arising from uncertainty regarding true roll diameter, particularly the case with coated rolls. It is common practice to use a so-called 'pi-tape' to obtain the effective diameter by measuring the circumference.

A check on the actual base inertia of the dynamometer is made either by means of a number of 'coast down' tests or by both accelerating and decelerating the rolls at an approximately constant rate, plotting motor torque and speed in both cases. The later method is the more accurate since it effectively eliminates the internal drag of the dynamometer motor.

It will be apparent that really accurate measurements of traction/braking force are by no means easy to achieve. Much depends on the quality of manufacture and consistency of performance of the machine, the accuracy of the basic measurements of torque, roll diameter and speed and the capability of the associated computer.

Limitations of the chassis dynamometer

Like all devices intended to *simulate* a phenomenon, the chassis dynamometer has its limitations, which are easily overlooked. It cannot be too strongly emphasized that a vehicle, restrained by elastic ties and delivering or receiving power through contact with a rotating drum, is *not* identical dynamically with the same vehicle in its normal state as a free body traversing a fixed surface. During motion at constant speed the differences are minimized, and largely arise from the absence of air flow and the limitation of simulated motion to a straight line.

Once acceleration and braking are involved, however, the vehicle motions in the two states are quite different. To give a simple example, a vehicle on the road is subjected to braking forces on all wheels, whether driven or not, and these give rise to a couple about the centre of gravity, which causes a transfer of load from the rear wheels to the front with consequent pitching of the body. On a chassis dynamometer, however, the braking force is applied only to the driven wheels, while the forward deceleration force acting through the centre of gravity is absent, and replaced by tension forces in the front-end vehicle restraints (braking on a rolling road tends to throw the vehicle backwards).

It will be clear that the pattern of forces acting on the vehicle, and its consequent motions, are quite different in the two cases. These differences make it very difficult to investigate vehicle ride, and some aspects of NVH testing, on the chassis dynamometer. A particular area in which the simulation differs most fundamentally from the 'real life' situation concerns all aspects of drive line oscillation, with its associated judder or 'shuffle'.

The authors have had occasion to study this problem in some detail and, while their analysis is too extensive to be repeated here, one or two of their conclusions may be of interest:

- the commonly adopted arrangement, whereby the roll inertia is made equal to that of the vehicle, gives a response to such disturbances as are induced by driveline oscillations, i.e. judder, that are significantly removed from on-road behaviour

- for reasonably accurate simulation, roll inertia should be at least $5\times$ vehicle inertia
- electronically simulated inertia is not effective in this instance: actual mass is necessary
- the test vehicle should be anchored as lightly and flexibly as possible: this is not an insignificant requirement, since it is desirable that the natural frequency of the vehicle on the restraint should be at the lower end of the range of frequencies, typically 5 to 10 Hz, that are of interest.

It is strongly recommended that investigations of vehicle behaviour under conditions of drive-line oscillation should proceed with caution if it is intended to involve running on a chassis dynamometer.

Summary

The subject of rolling roads and chassis dynamometers is a very extensive one, worthy of a textbook in its own right, though to the best of the authors' knowledge such a book has yet to be written. However, the testing techniques employed have a great deal in common with those applicable to engine testing dynamometers, and almost the whole of the present book is relevant.

In this chapter an attempt has been made to survey the whole range of apparatus that may be described by the general heading of rolling roads/chassis dynamometers, from the simple garage brake tester to complex and very costly climatic chambers and anechoic cells. The design of the associated test cells, vehicle restraint systems, driver communications and safety precautions have been discussed and problems of accuracy and calibration considered.

The *application* of these machines, at all but an elementary level, is very largely in the field of transient testing, the subject of Chapter 15. This deals with the theoretical basis for the planning and interpretation of much chassis dynamometer work: the road load equation.

Reference

1. *Tire Rolling Resistance*, Symposium, 122nd Meeting, Rubber Division, American Chem. Soc., 1982.

17 Tribology and engine testing

The science of tribology, launched as a separate discipline with a new name by the Jost Report of 1966, is concerned with lubrication and wear. Both are central to i.c. engine operation. It was pointed out some years ago that the difference between a brand new road vehicle and one that was totally worn out was the loss of perhaps 100 g of material at critical points in the mechanism. So far as friction is concerned the great majority of the power developed by an automobile engine is ultimately dissipated as friction. Even a large slow-speed diesel engine is unlikely to achieve a mechanical efficiency exceeding 85%, most of the losses being due to friction.

As evolution proceeds different aspects of engine lubrication and wear call for concentrated attention. Some problems, such as excessive bore wear and bore polishing in diesel engines, are largely overcome and sink into the background. Others emerge as user requirements change, operating conditions become more demanding or legislation sets new parameters. The following may be regarded as the main tribology-related problems of particular current concern:

- valve train wear
- control of lubricant consumption and its effect on emissions
- interactions between fuel and lubricant
- development of synthetic and 'fill for life' lubricants
- piston design, with particular reference to top land clearance, ring wear, oil control, distortion and reduction of friction.

The pursuit of higher efficiency and reduced friction losses is a continuing activity in all engine development departments.

Tribological test techniques

A feature of the new science has been the development of a great number of test techniques[1,2] that do not involve the running of complete engines. While these may be seen as outside the remit of the test engineer he should have a general awareness of them since they generally cost far less and are much quicker than engine or field tests. The cost of developing a new engine is enormous and it is prohibitively expensive to attempt to test all potential new materials, surface finishes or lubricants in the engine.

Table 17.1. *Test options in tribology*

Physical	Mechanical	Tensile strength, hardness, ductility, fatigue strength, shear stability, viscosity
	Chemical	Elemental composition, total base number (TBN), volatility, flash point, pour point, foaming, seal compatibility, water separation
Bench		Standard four-ball EP, four-ball wear, Falex, Timken, FZG gear scuffing
	General	Abrasion test, pin-on-disc, block-on-ring, reciprocating pin-on-plate
Engine	Fired	Examples include Peugeot TU3, Mercedes M102E, Sequence III, IV and V tests
	Motored	Examples include Toyota 3AU, P-VW 5106 cam/tappet wear, PSA XN-1 cam/tappet pitting, instrumented single cylinder engines
Field		Real production vehicles

Table 17.2. *Costs and benefits of test options*

Type of test	Duration	Cost ($)	Comments
Physical	Hours	50–500	Well controlled, repeatable but relates to specific properties, not performance
Bench	Days	500–5000	Well controlled, repeatable and can be related to practical performance
Engine	Weeks	15 000–40 000	Poor control and repeatability but close to practical performance
Field	Months	400 000 plus	Very little control but is the real thing

It is here that bench tests can reduce costs both by improving the fundamental understanding of friction and wear processes in the engine and by eliminating unpromising candidates at an early stage. Table 17.1 shows the range of options open, with a few references to standard engine tests, while Table 17.2 gives a rough indication of relative costs.

Physical tests

These are well-controlled, cheap and repeatable tests to assess the quality or nature of a product. Their relation to engine performance is remote but they can be a useful means of screening out definitely unpromising candidates and an aid to quality control. Their importance in the lubricant world may be judged by the number of such standard tests that must be passed for approval. Thus, for example, the UK Ministry of Defence Oil Specification for gasolene and diesel engine oils calls for the following bench tests:

- kinematic viscosity; apparent viscosity; shear stability
- corrosion inhibition; pour point; flash point
- foam tendency; foam stability; effect on synthetic rubber seals together with three standard engine tests, ref. 1.

Bench tests

A wide range of standard tests are written into standards such as those of the American Society for Testing and Materials (ASTM) and the Institute of Petroleum (IP). A well-known example, the four-ball test, dates from the 1930s and is still widely specified. The Timken test is equally venerable. In the absence of alternatives the machines for these tests have been used for lubricant and additive development but they are not very satisfactory since the test conditions they impose are remote from any to be found in a real engine.

Two machines show more promise. The first uses the pin-on-disc configuration and has been traditionally used for testing the wear properties of materials in both dry and lubricated conditions. The principal advantage of this configuration is the possibility of varying load intensity and sliding speed over a wide range. A disadvantage is the difficulty of controlling temperature, the governing factor in lubricant and additive performance. The reciprocating pin-on-plate machine meets this difficulty and has come into wide use for the simulation of such phenomena as valve train and piston ring wear or the scuffing of gear teeth.

One of the important factors that determines whether such simulations are of value is their ability to reproduce the wear and/or failure conditions experienced in the engine. A number of analytical and diagnostic tools are available to assist in this understanding, Table 17.3. An examination of specimens from the engine and from the test rig will determine whether the conditions are realistically reproduced.

Engine tests

Steady state engine dynamometer tests are of two main kinds; the fired engine and the motored engine. Fired tests aimed at assessing lubricant and material properties have the advantage that the test conditions can be close to those met in service. The major disadvantages are the cost and time involved and the difficulty, where measurements of such factors as surface wear, deposits, oil consumption and friction are concerned, of obtaining consistent results. The

Table 17.3.   *Surface examination techniques*

Visual and microscopic examination
Wear debris quantifying and analysis
Replication techniques
Surface roughness measurement
Metallographical sections and etching
Microhardness measurement
Scanning electron and scanning tunnelling electron microscopy
X-Ray, Auger and SIMS surface elemental analysis
Infra-red and Raman spectroscopy

measurements can be much affected by variations in engine build, components and even fuel.

Motored engine tests are used for assessing specific components such as valve trains. They are at some remove from real conditions but can give a good insight into local problems of wear and lubrication. By the successive removal of components from a motored engine a good estimate can be made of the contribution made to frictional losses by the different components, Fig. 17.1.

Appendix II lists some of the very numerous standard tests that are in constant use for assessment and routine confirmation of lubricant properties. The development, updating and monitoring of these standard tests, not to mention the cost in fuel, components and man-hours, represent a major burden to engine and lubricant manufacturers. A serious problem is that the standard engines specified soon become out of date and fail to represent current practice. See Table 17.4 as an example of what is involved in such standard tests.

Examples of engine tests involving tribology

Wear of engine camshafts with finger followers   Advances in combustion chamber design in spark ignition engines have made necessary the adoption of overhead camshafts with finger followers[3]. These tend to suffer from severe scuffing problems, resulting in the proliferation of standard tests, Table 17.4. One of the authors has been involved in a theoretical study which has led to a better understanding of the cause of the peculiar wear pattern observed on the follower, Fig.17.2a. This has been found to correlate with a factor entitled the 'energy pulse', a measure of the local intensity of energy dissipation in the contact, Fig. 17.2b. This was a case in which engine tests were linked with bench simulations.

Measurement of compression ring oil film thickness   A feature of many tribological situations is the extreme thinness of the oil films present, usually to be measured in microns and various techniques have been developed for

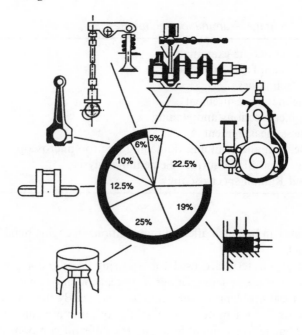

Figure 17.1. *Distribution of friction losses in an i.c. engine*

measuring these very small clearances[4]. Figure 17.3 shows measured oil film thicknesses between top ring and cylinder liner near top dead centre in a large turbocharged diesel engine. There is a wide variation in film thickness for the different strokes, and an effective breakdown during the firing stroke. It is notable that a thinner oil, SAE 10W/30, gives better protection than a thicker one. This was thought to be the result of higher oil flow to this critical area with the thinner oil and is confirmed by the generally lower rates of wear of aluminium from the piston and chromium from the ring surfaces, Fig. 17.4.

Wear measurements by irradiation Figure 17.5 shows wear measurement using an irradiated piston ring in an experimental single cylinder diesel engine and indicates a steady wear rate following initial running-in. In this test also wear rate was found to be sensitive to oil viscosity. This technique is perhaps the only one available for indicating very small increments of wear.

Analysis of friction losses

Figure 17.1 indicates the large contribution made to the friction losses in the engine by the piston and rings. Detail design changes can result in considerable economies, thus a piston design incorporating pads raised little more than one

Table 17.4. *Current low-temperature valve train wear tests*

Engine	Fired	Fired	Fired	Motored cyl. head
Fuel	Leaded	Unleaded	Unleaded	None
Power (kW)	100	47	67	N/A
Cam	Induction hardened iron	Cast iron	Cast iron	Cast iron
Follower	Induction hardened iron with chrome plate	Hard steel	Hard steel	Phosphated iron
Average wear (μm)	25	15	130	10
Duration (h)	175	100	288	200
Temperature (°C)	40–120	40–100°C	68–99°C	60°C

thousandth of an inch above the general piston skirt surface has been shown to give improvements in fuel consumption of nearly 4%, Fig. 17.6.

Bore polishing in diesel engines

This rather mysterious phenomenon has given a good deal of trouble. It is characterized by the wearing away of the initial honed finish in the bore and its replacement by a mirror-like surface that is so smooth that it fails to retain the lubricant, leading to serious surface damage. It is believed to be a mechanical polishing process in which hard particles of carbon generated by the combustion process are trapped in the top land of the piston and act as the polishing agent. It has been dealt with by reducing both width and length of top land clearance and, in some large engines, by fitting a hardened steel collar of reduced bore at the top of the liner.

Study of fuel and lubricant interactions

This is a major concern of lubricant manufacturers and involves the whole range of tribological test procedures: physical and bench tests and work on the test bed and in the field. A good example is work that has been done on the development of total-loss cylinder lubricants for large cross-head marine diesel engines burning high-sulphur content fuels. In these engines mean piston speed can be as low as 5 m/s, about the speed of a man riding a bicycle, a timed

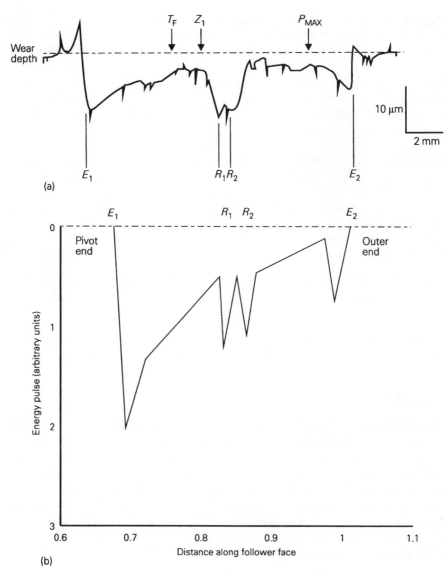

Figure 17.2. *Relation between wear pattern and energy pulse, finger type cam follower: (a) wear pattern on cam follower; (b) energy pulse*

supply of oil is injected through holes in the liner and expected bore wear rates are of the order of 0.01 mm per 1000 h.

Clearly test bed operation is scarcely practicable, on grounds of both size and time required, while information from engines in service at sea is both difficult to organize and slow to accumulate. Reference 5 describes the

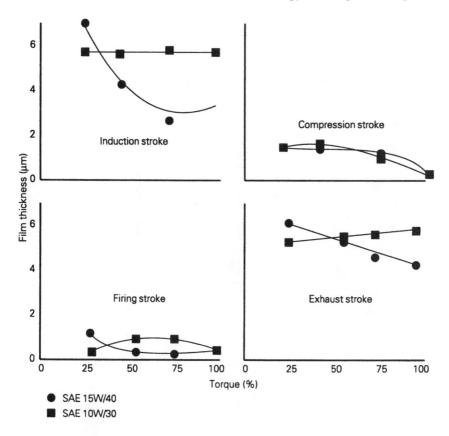

Figure 17.3. *Oil film thickness between ring and cylinder liner*

development of a successful laboratory technique for simulating the wear process. This makes use of a small specimen of piston ring material which is reciprocated in contact with a fixed plate of cylinder material. Temperatures up to 600°C may be achieved and the lubricant is supplied by a drip feed. The test procedure involved running-in followed by six hours at a plate temperature of 250°C. Measurements include wear by capacitance transducer, friction force and contact potential between fixed and moving specimen.

The effect of fuels with a high sulphur content was simulated by adding a trace of sulphuric acid (which is formed in the combustion process) to the oil. The test results showed the significance of total base number (TBN) of the lubricant on the suppression of wear, and correlated well with experience at sea.

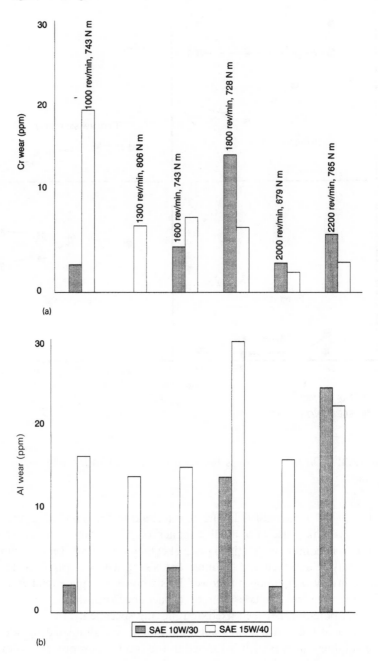

Figure 17.4. *Rates of wear, chromium from ring and aluminium from piston, engine as for Figure 17.3.*

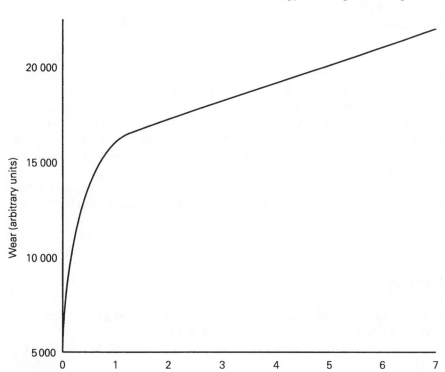

Figure 17.5. *Rate of piston ring wear, single cylinder diesel engine, using an irradiated ring and scintillation counter*

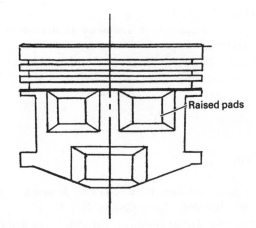

Figure 17.6. *Piston skirt modified to reduce friction ('Econoglide')*

Summary

These few examples will give some indication of the great range of tribological problems that may call for investigation under the general heading of engine testing. The point to be borne in mind by the test engineer who may be confronted with a problem of wear or friction is that it is unlikely to be solved on the test bed alone: there exists a vast range of specialized tests and techniques that eliminate a great deal of expensive engine testing and in many cases can clarify aspects of the problem that are simply not susceptible to investigation in the complex environment of a running engine.

For guidance in the tribological aspects of engine performance and testing the *Tribology Handbook* edited by M.J. Neale will be found valuable[6,7].

References

1. *Crankcase Oil Specification and Engine Test Manual* (1987) Paramins, Exxon Chemical, Technology Centre, Abingdon.
2. *Friction and Wear Devices* (1976) American Society of Lubrication Engineers.
3. Bell, J.C. (1989) Critical conditions for wear in pivoted-follower valve train systems, 3rd C.E.C. Symposium.
4. Moore, S.L. (1987), The effect of viscosity grade on piston ring wear and oil film thickness in two particular diesel engines, *Proc.I.Mech.E.* C184/87.
5. Davis, F.A. *et al.* (1993) The use of a laboratory wear simulation technique for the development of marine cylinder lubricants, 20th Int. Congress CIMAC, I.Mech.E., London.
6. Neale, M.J. (ed.) *Lubrication: A Tribology Handbook* (1993) Butterworth-Heinemann, Oxford.
7. Neale. M.J. (ed.) *Bearings: A Tribology Handbook* (1993) Butterworth-Heinemann, Oxford.

Further reading

Plint, M.A. and Alliston-Greiner, A.F. (1990) Routine engine tests: can we reduce their number. *Petroleum Review,* July 1990.
BS 2000 Parts 0 to 364, *Methods of Test for Petroleum and its Products.*

18 Data logging and computerized engine test cell control

One of the primary purposes, and sometimes the only purpose, of an engine test cell is to produce data. The collection, manipulation, display, storage and transmittal of this data should be one of the prime considerations in the design and operation of any test facility.

It is perhaps in the area of recording, processing and storage of data that the most revolutionary changes in the practice of engine testing have taken place. It should not be assumed that traditional methods should necessarily be rejected: they may still represent the best and most cost-effective solution in certain cases. In this chapter different approaches to the problem are described, leading to an account of the processes involved in drawing up a specification for a data acquisition system.

The traditional approach

Figure 18.1 shows a traditional 'test sheet', in this case showing the observations used in the preparation of Figs. 13.1 and 13.2 of Chapter 13. The sheet records the necessary information for plotting curves of specific fuel consumption, bmep, power output and air/fuel ratio of a gasolene engine at full throttle and constant speed, with a total of 13 test points. All the necessary information to permit a complete and unambiguous understanding of the test, long after the test engineer has forgotten all about it, are included.

In the experience of the authors, the same cannot necessarily be said of computer-acquired test data. It is a great temptation to leave the computer to 'get on with the job' without paying sufficient attention to the proper annotation of the information it has generated.

Such simple test procedures as the production of a power-speed curve for a rebuilt diesel engine can perfectly well be carried out with recording of the test data by hand on a pro-forma test sheet. There would be little justification for the installation of a computerized control and data acquisition system in the engine rebuild shop of a city bus company handling perhaps 20 engines per month. If an impressive 'birth certificate' for the rebuilt engine is required the test sheet may easily be transferred to a computer spread sheet and accompanied by a computer-generated performance curve.

TEST SHEET — I.C. ENGINES

DATE	CUSTOMER		UNIT NO.
DYNAMOMETER TORQUE ARM	205 mm		WORKS ORDER NO.

ENGINE	VARIABLE COMP.	BORE	85 mm	STROKE	82.5 mm	CYLINDERS	1	SWEPT. VOL.	468 cc	FUEL	SHELL 4-STAR	OIL	SHELL 20W30

BAROMETER	755 mmHg	AIR TEMP.	21 °C	AIR BOX SIZE	150 litre	ORIFICE DIA.	18.14 mm	FLOWMETER NO.		FUEL GAUGE 'O'

POWER:kW = $\dfrac{F_N\, n}{36{,}000}$ b.m.e.p. = $7.123\, F_N$ kN/m² FUEL Litre/Hour = $\dfrac{180}{t}$

NOTE: COMPRESSION RATIO 7 : 1

TACHO rev/min	COUNTER rev (2 x)	COUNTER sec	rev/min n	BRAKE LOAD F_N	POWER kW	b.m.e.p. kN/m²	t sec 50/cc	FUEL litre/hour	FUEL litre/kW·hr	EXH. °C	HEAD cmH₂O h_o	TEMP. °C T_A	VOL F/R l/s V_o	EFFY η VOL	MASS F/R g/s m_a	a/f RATIO	REMARKS
2000	780		2012	50.5	2.072	360	47.0	3.830	1.357	620	6.15	21	4.92	0.627	5.89	7.31	
	834		2002	67.0	3.726	477	50.0	3.600	0.966		6.15		4.92	0.630	5.89	7.85	
	891		1922	68.0	3.630	484	52.5	3.429	0.945	500	6.13		4.91	0.655	5.88	8.23	
	962		2025	71.0	3.993	506	57.0	3.158	0.791		6.10		4.90	0.620	5.87	8.92	
	1044		1989	74.0	4.089	527	63.0	2.857	0.699		6.05	21.5	4.88	0.629	5.84	9.81	
	1156		2010	74.5	4.160	530	69.0	2.609	0.627		6.00		4.86	0.620	5.82	10.71	
	1304		2019	74.5	4.178	530	77.5	2.323	0.556		5.97		4.85	0.616	5.80	11.99	
	1355		1995	74.5	4.129	530	81.5	2.209	0.535	675	5.93		4.83	0.621	5.79	12.58	
	1560		2035	69.0	3.900	491	92.0	1.957	0.502	685	5.93	19	4.83	0.609	5.79	14.20	
	1615		1988	65.2	3.601	464	97.5	1.846	0.513		5.87		4.81	0.620	5.76	14.98	
	1698		2008	60.0	3.347	427	101.5	1.773	0.530	673	5.95		4.84	0.618	5.80	15.70	
	1792		2010	52.0	2.903	370	107.0	1.682	0.580		6.03		4.88	0.623	5.84	16.66	
	1908		2000	35.0	1.944	249	114.5	1.572	0.809	682	6.03	19	4.88	0.626	5.84	17.83	

RICH → (top) ... WEAK (bottom)

DENSITY OF FUEL 0.75 kg/litre

Figure 18.1. Traditional 'Test Sheet'

Chart recorders

These represent a primitive stage in the process of automated data acquisition. Multi-pen recorders, having up to 12 channels giving continuous analogue records in various colours, still have a useful role, particularly in the testing of medium and large diesel engines where the performance of each cylinder unit may vary over time and call for individual adjustment. The clear and immediate indication of trends given by a chart recorder, and the ease with which the record may be annotated are not equalled by VDU based displays.

Computerized data recording with manual control

As the number of channels of information increases beyond about twelve and there is a need to calculate and record derived quantities such as power and fuel consumption the task of recording data by hand becomes tedious and prone to error.

An example, at a slightly higher level of complexity than the previous one, might be a small independent company producing specially-tuned versions of mass-produced engines for sports car use. The test cell will be required to produce detailed power curves and will also be used for development work.

Information to be recorded is likely to include:

- torque, speed and fuel consumption
- turbo-charger speed, temperatures and pressures
- several critical gas and component temperatures

The total required number of data channels could exceed 40. Typical equipment would be a PC-based data acquisition package with 16 pressure channels, 16 temperature channels, four channels for receiving pulse signals plus torque transducer signal processing and a facility for processing input from a fuel consumption meter.

All signals would be conditioned and linearized in the PC and displayed in tabular form. Proprietary data collection and signal conditioning cards are available with 8-12-, and 16-bit resolution and may be selected as appropriate to the accuracy and the significance of the signals to be processed.

No control, either of the test sequence or of the engine itself, is exercised by the PC. A 'snap shot' record of all channels may be taken and written to disc every time the operator presses a particular key of the test cell keyboard. Each such record will be time and date stamped with a 'time of test' figure derived from the PC's internal clock. At any time the operator is able to call up a power curve for any of these data sets.

It is entirely possible to adapt a PC data acquisition system at this level to give automatic test sequence control without large additional expenditure provided the dynamometer and engine controllers are designed to accept external commands.

However, if there is a wide variety of test sequences, not repeated very often, it can take longer to modify or programme the test sequence than to run the tests under manual control. At all levels of complexity the appropriate degree of sophistication in the management of the test and the recording of the data calls for careful consideration.

Computerized control and data acquisition

Fully computerized test cells are inevitably computer dependent: if the computer and its software are not functioning then the test bed may be effectively unusable. The degree of dependency is a function of the detailed design of the equipment; some designs will allow manual operation with the computer out of action but many will not.

In the following sections the various functions of the fully comprehensive test bed computer are considered and the many considerations to be taken into account in assessing and comparing such systems are indicated.

Data acquisition, signal conditioning, data manipulation and display, tabular and graphical

Certain tests, such as those required for engine emission homologation, may be pre-programmed in all details by the computer supplier, but normally the user is provided with a set of software tools to set up the required functions: data collection, labelling, scan rates, logging frequency and display of data.

There should be the facility to set a limited number of channels to higher scan rates than for others but it is possible to place too much emphasis on scan rates when comparing systems. In certain types of transient test logging rates of 1 kHz or more may be required but for many parameters, fluid temperatures for example, such scan rates are inappropriate and may even, as in the case of engine speed, give misleading 'snapshots' of the value being measured. Higher than necessary scan rates can also cause unnecessary problems in data storage.

'Real time' display of selected data, often in a graphical format chosen by the operator, is becoming more common as the power of the test bed computer increases. Such displays can be very useful when setting up control loops.

'Smart' devices and systems: communication and control

The main test bed computer may be linked to smart instrumentation, sending the necessary triggering and control signals and recording the formatted data produced by the device. Linking the test bed control computer with the many types of instrument available is not a trivial task and it is essential, when choosing a main computer, to check on the availability and compatibility of equipment and data transfer protocols.

Control of auxiliary services and systems

There are two strategies available for the control of test cell support services: they may either be controlled directly or by way of their own control systems. In the second case the set point command may either be set manually or 'sent' by the test bed control computer. If a particular system is required to have a change of state during a test sequence it is obviously desirable that the main computer should be able to communicate with the system controller.

Control devices supplied by such companies as Honeywell or Siemens are usually mounted in the main control cabinet and have a default display of current condition of the controlled parameter. Most of these 'catalogue' control devices are designed to meet the widest possible range of applications and may not be ideal for test bed use. In particular it may be difficult to optimize their response to very rapid changes in engine power output and to avoid major overshoots in such devices as temperature controllers.

It is possible to find instruments that employ complex cascade controllers but it is important, if unwanted system interactions are to be avoided, to embody them as part of an integrated strategy for managing test cell services and environment. An ability to accept set point commands from the test cell computer is a desirable feature.

Most test bed control computers have available auxiliary PID loops which may be used to control directly the auxiliary temperature and pressure regulating devices in the cell.

Editing and control of complex engine test sequences

Structure of test sequences

The typical modern engine test programme consists of a long succession of stages of speed and torque settings, together forming a complete test profile or 'sequence'. It is beyond the capacity of a human operator to run such test sequences accurately and repeatably.

Although the terminology used by test equipment suppliers may vary the principles involved in building up test profiles will be based on the same essential component instructions.

Test sequence editors range from those that require the operator to have knowledge of software code to those that present the operator with an interactive 'form fill' screen. The nature and content of the questions shown on the screen will depend on the underlying logic which determines how the answers are interpreted, and it is important that the user should understand this logic. Otherwise, there may be a risk of calling up combinations of control that the particular system cannot run, though it may not be able to indicate where the error lies. For example, if the control mode is set at throttle: position and dynamometer: speed (p.34) the editor will 'expect' instructions in terms of these parameters; it will not in this case be able to accept a throttle control instruction in terms of torque.

Each sequence will be made up of a series of stages, each either steady state, engine speed and torque constant, or transient, covering a move from one setting of speed and torque to another (of course speed and torque may also remain constant and other parameters, such as EGR or boost pressure, may be varied).

Editing of test sequences

In most cases the test sequence editors embodied in a suite of softwear designed for the control of engine test beds are based on a simple 'form fill' layout. The following elements are involved:

Modes of control

The various control modes for engine throttle and dynamometer are described in Chapter 3, p.32. Some analogue based control systems require engine speed and load to be brought down to some minimum value between each stage but digital controllers should be able to make a 'bumpless transfer' between stages.

Engine speed and torque or throttle position

Note that the way in which these parameters are set will depend on the control mode. A good sequence editor will present only viable options and will give a warning should, for example, a value of throttle opening be given when the mode requires a torque value.

Ramp rate

This is the acceleration required or the time specified for transition from one state to the next.

Duration or 'end condition' of stage

The duration of the stage may be defined in several ways:

- at a fixed time after the beginning of the stage
- on a chosen parameter reaching a specified value
- on reaching a specified logic condition, e.g. on completion of a fuel consumption measurement

Choice of next stage

At the completion of each stage (stage x) the editor will 'choose' the next stage. Typical instructions governing the choice are:

- run stage $x + 1$
- re-run stage x a total of y times (possible combinations of 'looped' stages may be quite complex and include 'nested' loops)
- choose next stage on basis of a particular analogue or digital state (conditional stage)

Events to take place during a stage

Examples are the triggering of ancillary events such as fuel consumption measurement or smoke density measurement. These may be programmed in the same way as duration or end condition, above.

Nominated alarm table

This is a set of alarm channels that are activated during the stage in which they are 'called'. The software and wired logic must prevent such programming from overriding 'global' alarms.

Monitoring of alarm signals

During a test the condition of both engine and facility will be monitored continuously so that the appropriate pre-programmed action may be triggered should one or more parameters fall outside a pre-set limit. Alarms are of two kinds: those safeguarding the equipment and those safeguarding the integrity of the test. The first type will unconditionally cause protective actions to be taken; it is a common safeguard to have the most critical factors, such as engine oil pressure and dynamometer water flow, as part of a 'hard wired' alarm and shut-down circuit that is separate from the computer-monitored transducers and alarm system.

Most engine test computer software makes available four levels of alarm for each data channel:

High level—shut down
High level—warning
Low level—warning
Low level—shut down

Such comprehensive warning arrangements are likely to be necessary on only a few channels; in most cases they will either be 'switched off' or assigned values outside the operating range. Some software permits alarms to be set 'stage specific'; this can be useful when, for example, particularly close control of fuel temperature is required during a consumption measurement the warning limits can be closed just for the critical period.

Most engine test software contains a rolling buffer that allows examination of the values of all monitored channels for a short time before any alarm shut down. Known as historic or 'dying seconds' logging, this buffer will also record the order in which the alarm channels have triggered, information of help in tracing the location of the prime cause of a failure.

Computerized Calibration procedures

Software routines will be provided to allow the operator to calibrate the various measuring devices and systems. Often these routines follow a 'form fill' format that requires the operator to type in confirmation of test signals in the correct order, permitting zero, span and intermediate values to be inserted in the calibration calculations. Such software should not only lead the operator through the calibration routines but should also store and print out the results in such a way as to meet basic quality control and certification procedures. The appropriate linearization and conversion procedures required to turn transducer signals into the correct engineering units will be built into the software.

Software for production test cells

Computers required to manage production test cells are required to perform a number of specialized tasks, of which examples follow.

Bar code labels are commonly attached to engines arriving for test: these will contain all the information necessary for the test stand software to select the required test procedure with appropriate values, and to create a test 'header' that will form the basis of the engine test certificate. Following the test, the test bed computer may print a bar code label which may include fault codes. The increasing use of 'smart chips' attached to the engine allows the computer to read all the engine information with the use of non-contact devices.

The computer software for production test beds must contain much more information about the end user and the production logic of the plant than would be embodied in the software for an R & D test facility. The identity codes for engines, the procedure for dealing with minor repairs, the pass and fail criteria, integration with production control computers and many other details must be built into the application software at the time of facility design.

Production test cells often require the operator to make visual checks and perhaps make some physical adjustment, such as setting an idling stop, in the course of an otherwise automatic test sequence. An operator may supervise several cells and so will need to be called to a particular unit when his intervention is required, either for normal or for abnormal events. Flashing signals and clear display messages calling for acknowledgement will be required.

Post acquisition data processing and reporting

Post processing and display of data is often carried out using specialized proprietary software packages that can read the buffered or stored information directly from the test cell computer; sometimes this software is built into the total software suite but it may be necessary to process the data with the computer off-line. It may be important for purchasers of data acquisition systems to check that the data from their system can be converted into non-specific format such as ASCII for importing into spreadsheets and databases.

General guidelines for the choice of engine test software

Worldwide, there are probably less than 15 proven suites of software that have been developed, and are supported, primarily for the control of R & D engine test beds.*

For many users, including research and quality audit engineers, these proprietary suites together with the range of engine test hardware they support will meet all their requirements; the choice will be a matter of subjective preference though the level of technical support promised is an important consideration. Some users may have limited requirements that do not make full

---

*Software for production engine test cells is always designed specifically for the customer while some production cells can be controlled using PLCs. There is thus a far greater range of non-specific sources of control software and hardware that can be built up into production test cell control systems.

use of the facilities offered. Other users may have so many specialized needs that there will be a need for bespoke software or for a secondary system such as a PLC to control these functions under the overall control of the main testbed computer; an example is production testing, where there will be a requirement to control engine conveying and docking, documentation and operator prompts, etc.

It is sensible for a prospective purchaser of a new system to compare its performance with that of the existing system, since the suppliers will probably have emphasized different aspects of the wide range of functions that the system is required to control. Particular care should be taken to check that the computer is capable of communicating with all the devices that make up the cell system.

The screen displays should be assessed for their relevance to the tasks to be undertaken, for the ease with which they may be adapted to the user's requirements, and for the clarity of the logic underlying the menu and display hierarchy. It is to be remembered that the almost infinite variety of display that is possible must be managed and used in a safe and auditable manner. Errors and misuse do not merely corrupt data; they can also lead to dangerous mechanical failures.

The established suppliers of engine test cell computer systems will be familiar with all the usual safety logic applicable to test cell control and will have included it in the design of their equipment and software. The purchaser should however investigate the programmed sequence of events in the case of modes of failure for which he may have a particular concern.

Examples are:

- power supply failure: does the throttle close or remain in the position at failure?
- shaft failure: can the system compare engine and dynamometer speed to detect failure?
- failure to achieve set point or test profile: can the system set alarms when set conditions have not been met?

Test sequences for special purpose test cells

Anechoic cell testing

The data acquisition requirements for anechoic testing are very specialized: the engine or transmission is considered only as a noise generator. Certain sequences may require some support services such as ventilation to be intermittently shut down to reduce background noise and the cell interlock system must be able to accommodate such requirements.

Endurance testing

These tests are very expensive to run and call for a high level of performance and reliability in the control and data logging systems. Duplication of transducers and careful running of cables to prevent deterioration by wear or heat over long periods should be considered. The amount of data generated over 1000 hours of cyclic running can be enormous so data filtering and compression is usually required.

Thermal shock and thermal cycling tests

These call for special programming arrangements, 'end condition' of stage being often defined by the attainment of a given temperature. There will often be the need to operate fluid control valves in strict sequence in order to direct engine coolant flows to or from chilled water buffer stores. Since this type of testing may be expected to induce a failure in some part of the engine system the failure monitoring system must be designed accordingly.

Combustion analysis

This calls for the collection of data, particularly cylinder pressures, at the highest possible speed. The data are subsequently processed to produce curves such as Fig. 13.6, p.235.

Ultra-fast recording

For some purposes, such as the analysis of the performance of electronic engine management systems, it may be necessary to record data at a speed at which accurate 'time stamping' of individual events becomes difficult. 'Time skew' can arise when the time taken to record all the values corresponding to an event taking place at a given instant is significant relative to the interval between successive events.

High speed data recording can produce unwanted detail, for instance by sensing cyclic variations in crankshaft speed if speed is sensed on the basis of a very short measuring interval. (Incidentally it is well to avoid the practice of recording cylinder pressures on a time base rather than against crank angle as cyclic speed variations will then distort the record).

Management of data; some general principles

It cannot be too strongly emphasized that present-day data acquisition systems are capable of storing vast amounts of information that may be totally irrelevant to the purposes of the test.

Mere acquisition of the data is the 'easy' part of the operation. The real skill is required in the post-processing of the information, the distillation of the significant results and the presentation of these at the right time to the right people in a form they will understand. This calls for an adequate management system for acceptance (rejection), archiving, statistical analysis and presentation of the information.

Computerized systems are no more inherently accurate, nor traceable, than paper systems. All data should have an 'audit trail' back to the calibration standards to prove that the information is accurate to the level required by the user. It is a prime responsibility of the test cell manager, see Chapter 19, to ensure that the installation, calibration and recording methods associated with the various transducers and instruments are in accordance with the maker's instructions. Bad data are still bad data, however rapidly recorded and skillfully presented.

If the 'customers' of the test facility are not aware of the capabilities of the data acquisition system they will not make proper use of it. It is essential that they should be given proper training in the use of the facility to its maximum advantage.

Host computers

Host computers can play an important role in a test facility, providing they are properly integrated and not so remote from the test area as to degenerate into mere archives. They may be the points at which post-processing of the data takes place. The host should be able to repeat on its screen the display from any of the cells connected to it and to display a summary of the status of all cells. This will allow it to play a supervisory role.

The relationship between the host computer and the subordinate computers in the test cells must be carefully planned and disciplined. Under no circumstances should the lower level units be allowed to communicate with each other except by way of the host. However the division of work between cell computer and host may vary widely. Individual cells should be able to function and store data locally if the host computer is 'off line'.

With modern communications it is possible for the host computer to be geographically remote; some suppliers will offer modem linking as part of a 'site support' contract. This ability to communicate with a remote computer raises problems of safety since it is possible for remote staff to cause machinery, that they cannot see, to operate.

Data security

The control room of an engine test cell is rarely the ideal environment for storing test data. Rules are required for the management and storage of data, whether on local disc or at the host computer and access to the various functions must be restricted if the integrity of data and system control is to be preserved.

In most large engine test facilities there are three levels of security:

Operator

At this level tests may be started, stopped and run only when the 'header information', which includes engine number, operator name and other key information, has been entered. Operator code may be verified. No alarms can be altered and no test schedules edited but there will be a text 'notepad' for operator comments, to be stored with the other data.

Supervisor

On entry of password this level will allow alteration of alarm levels and test schedules, also activation of calibration routines, in each case by way of a form fill routine and menu entry. It will also give access to fault finding routines such as the display of signal state tables.

Engineer

On entry of password this level is allowed access to de-bugging routines and language level editing.

A note on passwords

Passwords are keys that can be abused and misused. Military practice requires that passwords should be changed at random intervals and this is a sound principle. Such a requirement can be computer generated by the operating system.

The role of the computer in transient testing

Transient testing, as it is understood in the automotive industry, is covered in Chapter 15, and is very largely concerned with the *simulation* of road performance. The role of the computer in vehicle simulation is central, indeed such work has only been made possible by the availability of ever faster and more powerful processing systems. It will become normal practice to take actual or idealized vehicle road data and require the test cell computer to translate this into an equivalent programme of engine torque and speed which can be run as a repeatable simulation of the road data.

The process by which such a hierarchy of processors is controlled is still evolving; the graphical presentation of all the data generated is a current challenge to software designers and test engineers, and will become more demanding with the evolution of engine control, power train and on-board monitoring systems in vehicles.

Summary: drawing up the specification for a control and data acquisition system

First, and most important, decide what level of system is required: hand written, computerized with manual control, fully computerized, host computer.

Consider separately: test sequencing, engine and dynamometer control, data acquisition, post processing. Specify the requirements for each function before looking for a single unit that meets them all.

Ask the test cell users to list the channels needed for current and intended tests. Merely asking the users to give the number of channels they would like will lead to expensive over-estimates.

Remember that the measurement of some engine performance characteristics (cylinder pressure diagrams for example) is particularly expensive. A general cost consciousness at this stage can have a permanent effect on running costs.

Be aware of step changes in the cost of data acquisition equipment. Most systems will be made up of I/O boards with 8 or 16 channels. 33 channels will be considerably more expensive than 32. Apparently small increases in scan rates can also be expensive.

Consider possible problems: corruption of data by interference, to be minimized by careful cable layout and isolation of signal cables; effects of signal conditioning and transfer (e.g. multiplexing).

Devise disaster recovery procedure and back-up strategy

Check ability of selected equipment to receive and transmit signals to and from associated equipment. Special software drivers may have to be written and this work can be complex and expensive. Interfacing with such equipment as exhaust emissions analysers can be very difficult if not planned for when the system is designed.

Finally, the choice of equipment must be reviewed in the light of company policy and the medium-term development strategy for computerization. In particular, consider the extent to which your choice will tie you to a specific hardware platform and whether that platform is widely available and supported by companies other than the supplier. It is inevitable that modifications to the software in the light of experience will be necessary and the supplier should be asked for an indication of his charges for such modifications.

Further reading

BS ISO/IEC 2382 Parts 1 to 26 *Information Technology. Vocabulary.*

BS 4058 *Specification for Data Processing Flow Chart Symbols, Rules and Conventions.*

BS 4505 Part 1 and 2 *Digital Data Transmission.*

BS 6488 *Code of Practice for Configuration Management of Computer-based Systems.*

BS 6527 *Specification for Limits and Methods of Measurement of Radio Interference Characteristics of Information Technology Equipment.*

BS 7083 *Recommendations for the Accommodation and Operating Environment of Computer Equipment.*

BS 7153 *Guide for Computer System Configuration Diagram (Symbols and Conventions).*

19 The context of the test department: organization and communications, the log book and the technical report. The design of experiments

This chapter describes the organization of a typical test department, explaining how it should fit into the overall manufacturing operation. Recommended procedures are given for the planning, execution and reporting of test programmes.

In the experience of the authors there is always a temptation for the dedicated test engineer to follow up lines of investigation that may be of absorbing interest but which do not necessarily advance the particular project in hand at the time. One must always leave room for the occasional brilliant inspiration that is the stuff of scientific advance, but for most of the time effort must be directed firmly to the specified goal.

For the great majority of the working day the test department will be working for a 'customer', perhaps located within the same organization, perhaps an outside client, more rarely on a general brief to investigate an area of future interest. In each case the first essential is to understand the brief, which must always include an indication of time and cost.

Management structure of a test department

The executive in overall charge of a test department has three distinct management responsibilities:

- day-to-day operation of the test facility and the various supporting services
- technical control of the work carried out in the department
- processing, storage and dissemination of the data which is the only end product of the whole operation.

These tasks are not mutually exclusive and in a small test shop may be merged, although in all but the smallest department management of the first task, which includes the all-important responsibility for calibration and the general reliability and accuracy of instrumentation, should be kept distinct from management of the uses to which the facility is put.

Figure 19.1 shows the allocation of tasks in a fully articulated test department and indicates the various areas of overlapping responsibility, also the impact of 'quality management'. Periodical calibration of dynamometers and instruments and the maintenance of calibration records may be directly in the hands of the quality manager or carried out by the facilities department under his supervision; either way the procedure must be clearly laid down.

Computerized data logging has reduced the role of the test cell technician, who used to spend much of his time reading instruments and filling in log sheets. Pressure on costs has reduced the number of technicians, who now spend more of their time rigging and setting up engines, while the graduate test engineer tends to spend more of his time in the control room in the immediate consideration of the data being generated.

Figure 19.2 shows a management structure appropriate to the tasks shown in Fig. 19.1, which is self-explanatory but emphasizes the relatively complex nature of the engine testing operation and the wide variety of skills involved.

Health and safety matters have already been mentioned several times in connection with test cell operation. While everyone shown in Fig. 19.2 has responsibilities in this connection a particular burden rests on the shoulders of the test cell supervisor.

Planning, executing and reporting the test programme

It must never be forgotten that the sole purpose of research and development testing is to prove, or improve, the performance of the engine or component that is the subject of the test. The sole purpose of most other tests is to check that the performance of the engine or complete system conforms to some specified standard.

It follows that the data assembled during the test is only of value in so far as it contributes to the achievement of these purposes. It is only too easy, now that facilities are so readily available for the collection of colossal quantities of data, to regard the accumulation of data as an activity of value in its own right.

It is essential that the engineer responsible for the operation of a test facility should keep this very much in mind. In the experience of the authors nothing is more helpful in this respect than the discipline of keeping a proper log book and of writing properly structured reports.

The whole idea of a written log book may be considered by some to be old fashioned in an age in which the computer has taken over. Computers, however, are still not able to think as the engineer is compelled to think when he tries to write down, at the end of a day, what he has actually achieved during that day.

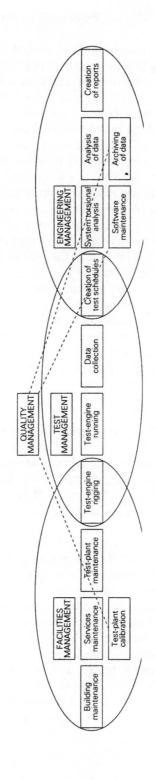

Figure 19.1. *Allocation of tasks in a test department*

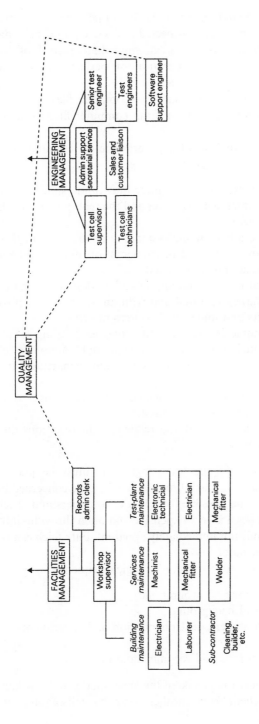

Figure 19.2. *Test department management structure*

The log book is also a vital record of all sorts of peripheral information on such matters as safety, maintenance, suspected faults in equipment or data recording and, last but most important, as an immediate record of 'hunches' and intuitions arising from a consideration of perhaps trivial anomalies and unexpected features of performance.

It is easy to lose track of information in computer files, particularly after the lapse of time, and cross referencing from computer files to the log book can help with this.

The execution and analysis of a programme of tests and experiments is a difficult art and involves a number of stages:

- First of all, the experimental engineer must understand the questions that his experiment is intended to answer and the requirements of the 'customer' who has asked them.
- There must be an adequate understanding of the relevant theory.
- The necessary apparatus and instrumentation must be assembled and, if necessary, designed and constructed.
- The experimental programme must itself be designed, with due regard to the levels of accuracy required and with an awareness of possible pitfalls, misleading results and undetected sources of error.
- The test programme is executed, the engineer keeping a close watch on progress (a technician, unless he is exceptional, cannot be relied upon to spot subtle and possibly highly significant departures from expected performance).
- The test data are reduced and presented in a suitable form to the 'customer' and to the level of accuracy required.
- The findings are summarized and related to the questions the programme was intended to answer.

Finally, the records of the test programme must be put together in a coherent form so that in a year's time, when everyone concerned has forgotten the details, it will still be possible for a reader to understand exactly what was done. Test programmes are very expensive and often throw up information the significance of which is not immediately apparent, but which can prove to be of great value at a later date.

Formal reports may follow a similar logical sequence:

- objective of experimental programme
- essential theoretical background
- description of equipment, instrumentation and experimental method
- calculations and results
- discussion, conclusions and recommendations.

In writing the report the profile of the 'customer' should be kept in mind. A customer who is a client from another company will require rather different treatment from one who is within the same organization. There will be common characteristics; the customer:

- will be a busy person who requires a clear answer to specific questions
- will probably not require a detailed account of the equipment - but will need a clear and accurate account of the instrumentation used and the experimental methods adopted
- will be concerned with the accuracy and reliability of the results
- must be convinced by an intelligent presentation that the problem has been understood and the correct answers given.

Statistical design of experiments

Throughout this book it has been the aim of the authors to give sufficient information concerning often complex subjects to enable the reader to deal with the more straightforward applications without reference to other sources. In the case of the present topic this is scarcely possible; all that can be done is to explain its significance and indicate where detailed guidance may be found.

The traditional rule of thumb in engineering development work has been *change one thing at a time*. It was thought that the best way to assess the effect of a design change was to keep everything else fixed. There are disadvantages:

- It is very slow.
- The information gained refers only to the one factor.
- Even if we have hit on the optimum value it may change when other factors are changed.

Consider a typical multi-variate development problem: optimizing the combustion system of a direct injection diesel engine, Fig. 19.3. The volume of the piston cavity is pre-determined by the compression ratio but there is room for manoeuvre in:

- overall diameter D
- radius r
- depth d
- angle of central cone θ
- angle of fuel jets ϕ
- height of injector h
- injection pressure p

not to mention other variables such as nozzle hole diameter and length and air swirl rate.

Clearly any attempt to optimize this design by changing one factor at a time is likely to over-run both the time and cost allocations and the question must be asked as to whether there is some better way.

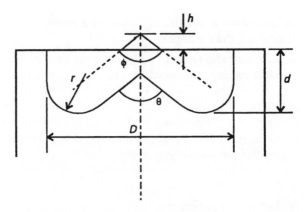

Figure 19.3.   *Diesel engine combustion system*

In fact techniques for dealing with multi-variate problems have been in existence for a good many years. They were developed mainly in the field of agriculture, where the close control of experimental conditions is not possible and where a single experiment can take a year. Published material from statisticians and biologists naturally tended to deal with examples from plant and animal breeding and did not appear to be immediately relevant to other disciplines.

The situation began to change in the 1980s, partly as a result of the work of Taguchi, who applied these methods in the field of quality control. These ideas were promoted by the American Supplier Institute and the term 'Taguchi methods', is often applied loosely to any industrial experiment having a statistical basis. A clear exposition of the technique as applied to the sort of work with which the engine developer is concerned is given by Grove and Davis[1] who have applied these methods in the Ford Motor Company.

Returning to the case of the diesel combustion chamber, suppose that we are taking a first look at the influence of the injector characteristics:

- included angle of fuel jets ϕ
- height of injector h
- injection pressure p

For our initial tests we choose two values, denoted by + and −, for each factor, the values spanning our best guess at the optimum value. We then run a series of tests in accordance with Table 19.1. This is known as an orthogonal array, the feature of which is that if we write + 1 and −1 for each entry the sum of the products of any two columns, taken row by row, is zero.

The final column shows the result of the tests in terms of the dependent variable, the specific fuel consumption. The chosen values were:

Table 19.1.   *Test programme for three parameters*

Run	ϕ	h	p	Specific consumption (g/kWh)
1	−	−	+	207
2	+	−	+	208
3	−	+	+	210
4	+	+	+	205
5	−	−	−	218
6	+	−	−	216
7	−	+	−	220
8	+	+	−	212

$$\phi- = 110°$$

$$\phi+ = 130°$$

$$h- = 2 \text{ mm}$$

$$h+ = 8 \text{ mm}$$

$$p- = 800 \text{ bar}$$

$$p+ = 1200 \text{ bar}$$

One way of presenting these results is shown in Fig. 19.4 in which each of the four pairs of results is plotted against each factor in turn. We can draw certain conclusions from these plots. It is clear that injection pressure is the major influence on fuel consumption and that there is some interaction between the three factors. There are various statistical procedures that can extract more information and guidance as to how the experimental programme should continue.

The first step is to calculate the *main effect* of each factor. Line *B* of Fig. 19.4(a) shows that in this case the effect of changing the injection pressure from $p-$ to $p+$ is to reduce the specific consumption from 212 to 205 g/kWh. The main effect is defined as one half this change or -3.5. The average main effect for the four pairs of tests is shown in Table 19.2.

A further characteristic of importance concerns the degree of interaction between the various factors. The interaction between ϕ and p is known as the $\phi \times p$ interaction. Referring to Fig. 19.4(a), it is defined as one half the difference between the effect of p on the specific consumption for $\phi+$ and the effect for $\phi-$, or

$$1/2[(205 - 202) - (210 - 220)] = +1.5$$

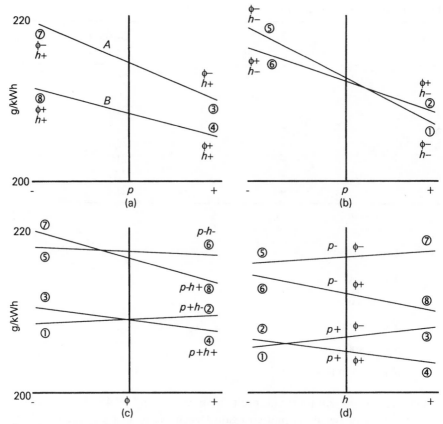

Figure 19.4.   *Relation between parameters showing main effects*

It would be zero if the lines in Fig. 19.4(a) were parallel. It may be shown that the inverse interaction $p \times \phi$, has the same value.

It may be shown that the sign of an interaction, $+$ or $-$, is obtained by multiplying together the signs of the two factors concerned. Table 19.2 shows the calculated values of the three interactions in our example. It will be observed that the strongest interaction, a negative one, is between h and ϕ. This makes sense, since an increase in h coupled with a decrease in ϕ will tend to direct the fuel jet to the same point in the toroidal cavity of the combustion chamber.

This same technique for planning a series of tests may be applied to any number of factors. Thus an orthogonal table may be constructed for all seven factors in Fig. 19.3 and main effects and interactions calculated. The results will identify the significant factors and indicate the direction in which to move for an optimum result. Tests run on these lines tend to yield far more information

Table 19.2. *Main effects and interactions*

Run	ϕ	h	p	$\phi \times h$	$\phi \times p$	$h \times p$
1	−	−	+	+	−	−
2	+	−	+	−	+	−
3	−	+	+	−	−	+
4	+	+	+	+	+	+
5	−	−	−	+	+	+
6	+	−	−	−	−	+
7	−	+	−	−	+	−
8	+	+	−	+	−	−
Main effect				Interaction		
	−1.75	−0.25	−4.5	−1.5	+0.75	+0.25

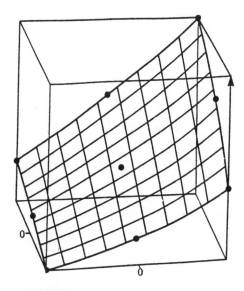

Figure 19.5. *Engine map, torque, speed and fuel consumption*

than simple 'one variable at a time' experiments. More advanced statistical analysis, beyond the scope of the present work, identifies which effects are genuine and which the result of random variation.

A more elaborate version of the same method uses factors at three levels, +, 0 and −. This is a particularly valuable technique for such tasks as the mapping of engine characteristics, since it permits the derivation of the coefficients of quadratic equations that describe the surface profile of the characteristic. Figure 19.5, reproduced from ref. 1, shows a part of an engine map relating speed, torque and fuel consumption. The technique involves the choice of three values, equally spaced, of each factor. These are indicated in the figure.

It is hoped that the reader will gain some impression of the value of these methods from the above, necessarily brief, treatment.

Summary

The management structure of a test department is described and the planning, execution and reporting of test programmes are discussed. The statistical design of experiments is briefly introduced.

Reference

1. Grove, D.M. and Davis, T.P. (1992) *Engineering Quality and Experimental Design,* Longmans, London.

20 The pursuit and definition of accuracy: statistical analysis of test results

This chapter is perhaps the most important, as well as the most difficult, in the book. Modern instrumentation and data logging have tended to obscure questions of accuracy, and to give an illusion of precision to experimental results that is often totally unjustified. This chapter points out the dangers and calls for a much greater awareness of the problem.

It is a statement of the obvious that the purpose of engine testing is to produce information, and that the value of that information depends critically on its accuracy. This presents the biggest challenge facing the experimenter, who in addition to a complete understanding of the engine under test must master the following skills:

- experience in the correct use of instruments
- knowledge of methods of calibration and an awareness of the different kinds of error to which instruments are subject
- a critical understanding of the relative merits and limitations of different methods of measurement and their applicability to different experimental situations
- an understanding of the differences between true and observed values of experimental quantities.

This is a very wide range of skills, only to be acquired by experience. The first essential is to acquire as a habit of mind a sceptical attitude to all experimental observations: all instruments tend to be liars.

The worst offender here is the digital readout, which gives an illusion of accuracy which may be, and usually is, totally unjustified. The temptation to believe that a reading of say, 97.12 is to be relied upon to the second decimal place is very strong, but in the absence of convincing proof must be resisted. An analogue indicator connected to a thermocouple cannot in most cases be read to closer than 1°C and the act of reading the indicator is likely to bring to mind the many sources of inaccuracy in the whole temperature measuring system.

Replace the analogue indicator by a digital instrument reading to 0.1°C and it is easy to forget that all the sources of error are still present: the fact that the read-out can now discriminate to ±0.05°C does *not* mean that the overall accuracy of the reading is within comparable limits.

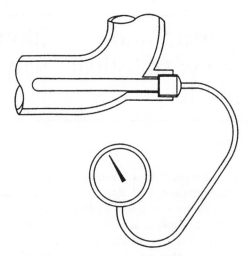

Figure 20.1.   *Measurement of exhaust temperature by vapour pressure thermometer*

Example: measurement of exhaust temperature of a diesel engine

Temperature measurements are some of the most difficult to make accurately, and an example illustrates many of the pitfalls encountered in the pursuit of accuracy[1]. Figure 20.1 shows a typical situation: the determination of the exhaust temperature of a diesel engine. The instrument chosen is a vapour pressure thermometer, such as is suitable for temperatures of up to 600°C. It comprises a steel bulb, immersed in a gas of which the temperature is to be measured, and connected by a long tube to a Bourdon gauge which senses the vapour pressure but is calibrated in temperature.

Let us consider the various errors to which this system is subject.

Sensing errors are associated with the interface between the system on which the measurements are to be made and the instruments responsible for those measurements. In the present case there are a number of sources of sensing error. In the first place, the bulb of the temperature indicator can 'see' the walls of the exhaust pipe, and these are inevitably at a lower temperature than that of the gas flowing in the pipe. It follows that the temperature of the bulb must be less than the temperature of the gas. This error can be reduced but not eliminated by shielding the bulb or by employing a 'suction pyrometer'. A further source of error arises from heat conduction from the bulb to the support, as a result of which there is a continuous flow of heat from the exhaust gas to the bulb and no equality of temperature between them is possible.

A more intractable sensing error arises from the circumstances that the flow of gas in the exhaust pipe is constant neither in pressure, velocity nor temperature. Pulses of gas, originating at the opening of the exhaust valves in individual cylinders, alternate with periods of slower flow, while the exhaust will also be to some extent diluted by scavenge air carried over from the inlet. The thermometer bulb is thus required to average the temperature of a flow that is highly variable both in velocity and in temperature, and it is unlikely in the extreme that the actual reading will represent a true average.

A more subtle error arises from the nature of exhaust gas. Combustion will have taken place, resulting in the creation of the exhaust gas from a mixture of air and fuel, perhaps only a few hundredths of a second before the attempt is made to measure its temperature. This combustion may be still incomplete, the effects of dissociation arising during the combustion process may not have worked themselves out, and it is even possible that the distribution of energy between the different modes of vibration of the molecules of exhaust gas will not have reached its equilibrium value; as a consequence it may not be possible even in principle to define the exhaust temperature exactly.

We can deal with some of these sensing errors by replacing the steel bulb of Fig. 20.1 by a slender thermocouple surrounded by several concentric screens, Fig. 20.2 and further improve accuracy by the use of a suction pyrometer, Fig. 20.3, in which a sample of the exhaust gas is drawn past the sensor at uniform velocity by external suction. This deals with the 'averaging' problem but still leaves the question regarding the definition of the exhaust temperature open.

Our example includes a Bourdon type pressure gauge and these are particularly prone to a variety of instrument errors.

Two of the commonest are zero error and calibration error. The zero error is present if the pointer does not return precisely to the zero graduation when the gauge is subjected to zero or atmospheric pressure.

Calibration errors are of two forms: a regular disproportion between the instrument indication and the true value of the measured quantity, and errors that vary in a non-linear manner with the measured quantity. This kind of fault

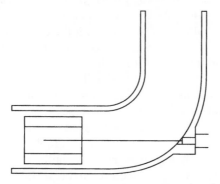

Figure 20.2. *Measurement of exhaust temperature by shielded thermocouple*

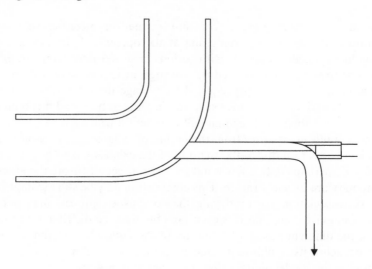

Figure 20.3.   *Measurement of exhaust temperature by suction pyrometer*

may be eliminated or allowed for by calibrating the instrument, in the case of a pressure gauge by means of a dead weight tester. These are examples of systematic errors.

In addition the pressure gauge may suffer from random errors arising from friction and backlash in the mechanism. These errors affect the repeatability of the readings.

The sensitivity of an instrument may be defined as the smallest change in applied signal that may be detected; in the case of a pressure gauge it is affected particularly by friction and backlash in the mechanism. The precision of an instrument is defined in terms of the smallest difference in reading that may be observed. Typically, it is possible to estimate readings to within 1/10 of the space between graduations, provided the reading is steady, but if it is necessary to average a fluctuating reading the precision may be much reduced.

Finally one must consider the effect of installation errors. In the present case these may arise if the bulb is not inserted with the correct depth of immersion in the exhaust gas or, as is quite often the case, if it is installed in a pocket and is not subjected to the full flow of the exhaust gas.

A consideration of this catalogue of possible errors will make it clear that it is unlikely that the reading of the indicator will reflect with any degree of exactness the temperature of the gas in the pipe. It is possible to analyse the various sources of error likely to affect any given experimental measurement in this way, and while some measurements, for example of lengths and weight, require a less complex analysis, others, notably readings of inherently unsteady properties such as flow velocity, require to be treated with scepticism. A hallmark of the experienced experimenter is that, as a matter of habit and training, he questions the accuracy and credibility of every experimental observation.

A similar critical analysis should be made of all instrumentation. This example has been dealt with in some detail to illustrate the large number of factors that must be taken into account.

Some general principles

(1) Cumulative measurements are generally more accurate than rate measurements. Examples are the measurement of speed by counting revolutions over a period of time and the measurement of fuel consumption by recording the time taken to consume a given volume or mass.

(2) Be wary of instantaneous readings. Few processes are perfectly steady. This is particularly true of flow phenomena, as will be clear if an attempt is made to observe pressure or velocity head in a nominally steady gas flow. A water manometer fluctuates continually, and at a range of frequencies depending on the scale of the various irregularities in the flow. No two successive power cycles in a spark ignition engine are identical. If it is necessary to take a reading instantaneously it is better to take a number in quick succession and average them.

(3) A related problem: 'time slope effects'. If one is making a number of different observations over a period of nominally steady state operation, make sure that there is no drift in performance over this period.

(4) The closer one can come to an 'absolute' method of measurement, e.g. pressure by water or mercury manometer or dead weight tester; force and mass by dead weights (plus knowledge the of local value of g)* the less the likelihood of error.

(5) In instrumentation, as in engineering generally, simplicity is a virtue. Each elaboration is also a potential source of error.

Definition of terms relating to accuracy

This is a particularly 'grey' area[1-3]. The statement 'this instrument is accurate to $\pm 1\%$' is in fact entirely meaningless in the absence of further definition.

---

* See p.164. Variations in the value of g, mainly a function of latitude, can give rise to errors in the measurement of *force* by *mass* (dead weights) or *mass* by *force* (spring balance). g departs from its nominal value, $9.81 \, \text{m/s}^2$, by $+0.2\%$ at the poles and -0.3% at the equator.

Table 20.1.   *Accuracy: definitions*

Error	The difference between the value of a measurement as indicated by an instrument and the absolute or true value
Sensing error	An error arising as a consequence of the failure of an instrument to sense the true value of the quantity being measured
Systematic error	An error to which all the readings made by a given instrument are subject; examples of systematic errors are zero errors, calibration errors and non-linearity
Random error	Errors of an unpredictable kind. Random errors are due to such causes as friction and backlash in mechanisms
Observer error	Errors due to the failure of the observer to read the instrument correctly, or to record what he has observed correctly
Repeatability	A measure of the random scatter of successive readings of the same quantity
Sensitivity	The smallest change in the quantity being measured that can be detected by an instrument
Precision	The smallest difference in instrument reading that it is possible to observe
Average error	Take a large number of readings of a particular quantity, average these readings to give a mean value and calculate the difference between each reading and the mean. The average of these differences is the average error; roughly half the readings will differ from the mean by more than the average error and roughly half by less than the average error

Table 20.1 defines various terms that are used, often incorrectly, in any discussion of accuracy.

Statements regarding accuracy: a critical examination

We must distinguish between the accuracy of an observation and the claimed accuracy of the instrument that is used to make the observation. Our starting point must be the *true value of the quantity to be measured*, Fig. 20.4(a).

The following are a few examples of such quantities associated with engine testing:

- fuel flow rate
- air flow rate
- output torque

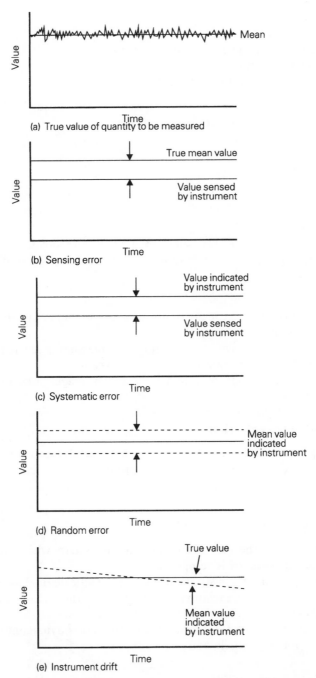

Figure 20.4. *Various types of error: (a) true value of quantity to be measured; (b) sensing error; (c) systematic error; (d) random error; (e) instrument drift*

- output speed
- pressures in the engine cylinder
- pressures in the inlet and exhaust tracts
- gas temperatures
- analysis of exhaust composition.

By their nature, none of these quantities is perfectly steady, and decisions as to the sampling period are critical, see above.

Sensing errors

These are particularly difficult to evaluate and can really only be assessed by a systematic comparison of the results of different methods of measuring the same quantity.

Sensing errors can be very substantial. In the case discussed above regarding measurement of exhaust temperature the authors have observed errors as great as 70°C, Fig. 20.4(b).

Typical sources of sensing error include:

(1) mismatch between temperature of a gas or liquid stream and the temperature of the sensor;
(2) mismatch between true variation of pressure and that sensed by a transducer at the end of a connecting passage;
(3) failure of a transducer to give a true average value of a fluctuating quantity;
(4) air and vapour present in fuel systems;
(5) time lag of sensors under transient conditions.

Note that the accuracy claimed by its manufacturer does not usually include an allowance for sensing errors. These are the responsibility of the user.

Systematic instrument errors, Fig. 20.4(c)

Typical systematic errors include:

(1) zero errors: the instrument does not read zero when the value of the quantity observed is zero;
(2) scaling errors: the instrument reads systematically high or low;
(3) non linearities: the relation between the true value of the quantity and the indicated value is not exactly in proportion;
(4) dimensional errors: for example the length of a dynamometer torque arm may not be precisely correct.

Random instrument errors

Random errors, Fig. 20.4(d), include:

(1) effects of stiction in mechanical linkages;
(2) effects of friction, for example in dynamometer trunnion bearings;
(3) effects of vibration or its absence;
(4) analogue to digital conversion,

Instrument drift

Instrument drift, Fig. 20.4(e), is a slow change in the calibration of the instrument as the result of:

(1) changes in instrument temperature;
(2) effects of vibration and fatigue;
(3) fouling of the sensor, blocking of passages, etc.;
(4) inherent long term lack of stability.

It will be clear that it is not easy to give a realistic figure for the accuracy of an observation.

Instrumental accuracy: manufacturers' claims

With these considerations in mind it is possible to look more critically at the statement 'the instrument is accurate to within ±1% of full scale reading'. Several different interpretations are possible:

(1) *No reading will differ from the true value by more than 1% of full scale.* This implies that the sum of the systematic errors of the instrument, plus the largest random error to be expected, will not exceed this limit.
(2) *The average error is not more than 1% of full scale.* See the definition of average error in Table 20.1. This implies that about half of the readings of the instrument will differ from the average by less than 1% full scale and half by more than this amount. However the definition implies that systematic errors are negligible: we are only looking at random errors.

Neither of these definitions is satisfactory. In fact the question can only be dealt with satisfactorily on the basis of the mathematical theory of errors.

Uncertainty

While it is seldom mentioned in statements of the accuracy of a particular instrument or measurement the concept of uncertainty is central to any meaningful discussion of accuracy. Uncertainty is a property of a measurement, not of an instrument:

The uncertainty of a measurement is defined as the range within which the true value is likely to lie, at a stated level of probability.

The level of probability, also known as the confidence level, most often used in industry is 95%. If the confidence level is 95% there is a 19 to 1 chance that a single measurement differs from the true value by less than the uncertainty and one chance in 20 that it lies outside these limits.

If we make a very large number of measurements of the same quantity and plot the number of measurements lying within successive intervals, we shall probably obtain a distribution of the form sketched in Fig. 20.5. The corresponding theoretical curve is known as a normal or Gaussian distribution, Fig. 20.6. This curve is derived from first principles on the assumption that the value of any 'event' or measurement is the result of a large number of independent causes (random sources of error).

The normal distribution has a number of properties shown on Fig. 20.6:

- The *mean value* is simply the average value of all the measurements.
- The *deviation* of any given measurement is the difference between that measurement and the mean value.
- The *variance* σ^2 (sigma)^2 is equal to the sum of the squares of all the individual deviations divided by the number of observations.
- The *standard deviation* σ is the square root of the variance.

The standard deviation characterizes the degree of 'scatter' in the measurements and has a number of important properties. In particular the 95% confidence level corresponds to a value $\sigma = 1.96$. 95% of the measurements will lie within these limits and the remaining 5% in the 'tails' at each end of the distribution.

In many cases the 'accuracy' of an instrument as quoted merely describes the average value of the deviation, i.e. if a large number of measurements are made about half will differ from the true or mean value by more than this amount and about half by less. Mean deviation $= 0.8\sigma$ approximately.

However this treatment only deals with the random errors: the systematic errors still remain. To give a simple example consider the usual procedure for checking the calibration of a dynamometer torque transducer. A calibration arm, length 1.00 m, carries a knife-edge assembly to which a dead weight of 10.00 kg is applied. The load is applied and removed 20 times and the amplifier output recorded. This is found to range from 4.935 V to 4.982 V with a mean value 4.9602 V.

At first sight we could feel a considerable degree of confidence in this mean value and derive a calibration constant:

$$k = \frac{4.9602}{10.00 \times 9.81 \times 1.00} = 0.050563 \text{ V/N m}$$

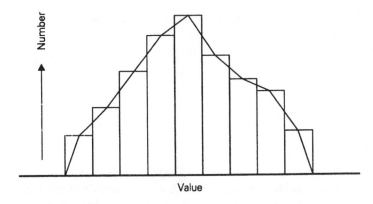

Figure 20.5. *A frequency distribution*

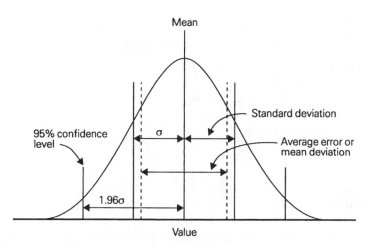

Figure 20.6. *Normal or Gaussian distribution*

The 95% confidence limit for a single torque reading may be derived from the 20 amplifier output readings and, for the limiting values assumed, would probably be about ±0.024 V, or ±0.48%, an acceptable value.

There are, however, four possible sources of systematic error:

- the local value of g may not be exactly $9.81 \, \text{m/s}^2$ (see p.164).
- the mass of the dead weight may not be exactly 10 kg
- the length of the calibration arm may not be exactly 1.00 m
- the voltmeter used may have its own error.

In fact none of these conditions can ever be fulfilled with absolute exactness.

We must widen our 95% confidence band to take into account these probably unknown errors.

This leads straight on to our next topic.

Traceability

In a properly organized test department careful thought must be given to this matter. In many small departments the attention given to periodic recalibration is minimal. In many more, while some effort may be made to check calibrations from time to time, the fact that the internal standards used for these checks may have their own errors is overlooked. To be confident in the validity of one's own calibrations these standards must themselves be checked from time to time.

Traceability refers to this process. The 'traceability ladder' for the dead weight referred to in the above example might look like this:

• the International kilogram (Paris)
• the British copy (NPL)
• national secondary standards (NPL)
• local standard weight (portable)
• our own standard weight.

Combination of errors

Most derived quantities of interest to the experimenter are the product of several measurements, each of them subject to error. Consider for example the measurement of specific fuel consumption of an engine. The various factors involved and typical 95% confidence limits are as follows:

• torque ±0.5%
• number of revolutions to consume measured mass of fuel ±0.25%
• actual mass of fuel ±0.3%

The theory of errors indicates that in such a case, in which each factor is involved to the power 1, the confidence limit of the result is equal to the square root of the sum of the squares of the various factors, i.e. in the present case the 95% confidence limit for the calculated specific fuel consumption is:

$$\sqrt{0.5^2 + 0.25^2 + 0.3^2} = \pm 0.58\%$$

The number of significant figures to be quoted

One of the commonest errors in reporting the results of experimental work is to quote a number of significant figures some of which are totally meaningless – the *illusion of accuracy* once more. This temptation has been vastly increased by the now universal employment of digital read-outs.

Consider a measurement of specific fuel consumption, readings as follows:

engine torque   110.3 N m
number of revolutions   6753
mass of fuel   0.25 kg

specific consumption equals

$$\frac{3.6 \times 10^6 \times 0.25}{2 \times \pi \times 6753 \times 110.3} = 0.1923 \text{ kg/kWh}$$

Let us assume that each of the three observations has the 95% confidence limit specified in the previous paragraph. Then the specific fuel consumption has a confidence limit of $\pm 0.58\%$, i.e. between 0.1912 and 0.1934 kg/kWh. Clearly it is meaningless to quote the specific fuel consumption to closer than 0.192 kg/kWh.

Particularly in development work where the aim is often to detect small improvements this question must be kept in mind.

Absolute and relative accuracy

A topic linked with the last one. In a great deal of engine test work – perhaps the majority – we are interested in measuring relative changes and the effect of modifications. The absolute values of a parameter before and after the modification may be of less importance; it may not be necessary to concern ourselves with the absolute accuracy and traceability of the instrumentation. Sensitivity and precision may be of much greater importance.

The cost of accuracy: a final consideration

An engineer with responsibility for the choice of instrumentation should be aware of the danger of mismatch between what he thinks he requires and what is actually necessary for an adequate job to be done. The cost of an instrument to perform a particular task can vary by an order of magnitude; in most cases the main variable is the level of accuracy offered and this should be compared with the accuracy that is really necessary. It would be absurd to use a thermocouple with a high-precision indicator to monitor engine oil temperature when a simple dial type vapour pressure thermometer would be adequate.

Summary

The degree of understanding of the subject of accuracy is perhaps the main criterion by which the professional quality of a test engineer should be judged. A proper understanding calls for a wide range of knowledge and the principal elements of this are described, with a special warning as to the dangers of digital read-outs. An example of temperature measurement illustrates the problems and terms relating to accuracy are defined.

This leads to a discussion of the meaning of the term as used by instrument manufacturers, and different types of error are identified. The mathematical basis of the concept of uncertainty is given and the method of dealing with combined errors is described.

References

1. Hayward, A.T.J. (1977) *Repeatability and Accuracy,* Mechanical Engineering Publications, London.
2. Dietrich, C.F. (1973) *Uncertainty, Calibration and Probability,* Adam Hilger, London.
3. Campion, P.J. *et al.* (1973) *A Code of Practice for the Detailed Statement of Accuracy,* HMSO, London.

Further reading

BS 4889 *Method for Specifying the Performance of Electrical and Electronic Measuring Equipment.*
BS 5233 *Glossary of Terms used in Metrology*
BS 5497 Part 1 *Guide for the Determination of Repeatability and Reproducibility for a Standard Test Method by Inter-laboratory Tests.*
BS 7118 Part 1 and 2 *Linear and Non-linear Calibration Relationships.*

Appendix I   Notes on the choice of instruments and transducers

It is not the intention to attempt a critical study of the vast range of instrumentation that may be of service in engine testing. The literature of the subject is very extensive and a small selection is given below. The purpose is rather to draw the attention of the reader to the range of choices available and to set down some of the factors that should be taken into account in making a choice. Only standard measurements, common to almost every test, are considered.

Table AI.1 lists the various measurements. Within each category the methods are listed in approximately ascending order of cost. Some comments follow.

General comments

Many of the simpler instruments, e.g. spring balances, manometers, Bourdon gauges, liquid-in-glass thermometers, cannot be integrated with data logging systems. This may or may not be important.

Accuracy costs money. Most transducer manufacturers supply instruments to several levels of accuracy with steeply increasing cost, and over specification should be avoided. Accuracy without repeatability is of little use.

It is cheaper to buy a stock item rather than to specify special features.

Some transducers, particularly force and pressure, can be destroyed by overload. Ensure adequate capacity or overload protection.

Temperature compensation may be important.

Always read the maker's catalogue with care, taking particular note of accuracies, overload capacity, fatigue life.

Time interval

Time itself can be measured with the greatest ease to an accuracy much greater than is necessary in engine testing, but the precise location of events in time is very much more difficult[1]. When linked events very close together in time are sensed in different ways it can be very difficult to establish the exact order or to detect simultaneity.

Table AI.1.   *Instrumentation and transducers for frequently required measurements*

Measurement	Principal applications	Method
Time interval	Rotational speed	Tachometer Single impulse trigger Starter ring gear Shaft encoder
Force, quasi-static	Dynamometer torque	Dead weights + spring balance Hydraulic load cell Strain gauge transducer
Force, cyclic	Stress and bearing load investigations	Strain gauge transducer Piezo-electric transducer
Pressure, quasi-static	Flow systems: lubricant, fuel, water, pressure charge, exhaust	Liquid manometer Bourdon gauge Strain gauge transducer
Pressure, cyclic	In-cylinder, inlet, exhaust events. Fuel injection	Strain gauge transducer Capacitative transducer Piezo-electric transducer
Position	Throttle and other controls	Mechanical linkage and pointer, counter LVDT transducer Shaft encoder Stepper motor
Displacement, cyclic	Valve lift, injection needle lift	Inductive transducer Hall effect transducer Capacitative transducer
Acceleration	Engine balancing, NVH	Strain gauge accelerometer Piezo-electric accelerometer
Temperature	Cooling water, lubricant, inlet air, exhaust, in-cylinder, mechanical components	Liquid-in-glass Vapour pressure Liquid-in-steel Thermocouple Thermistor Electrical resistance Optical pyrometer Suction pyrometer

In general a tachometer should only be used for speed measurement when the reading is not to be used as one component in calculating another quantity, e.g. power = torque × rev/min × constant. It is better practice to count the revolutions.

A single impulse trigger, such as a pin in the rim of the flywheel, is satisfactory for counting purposes but not ideal for locating t.d.c., see Chapter 12 for a discussion of the problems involved. Multiple impulses picked up from a starter ring gear may be acceptable. Many dynamometers are fitted with a 60-tooth wheel as standard. For precise indications down to a fraction of a degree an encoder is necessary, but arrangements to drive it may be difficult, while if as is usually the case drive is taken from the non-flywheel end of the crankshaft torsional effects in the crankshaft may lead to errors which may differ from cylinder to cylinder.

Force, quasi-static

The combination of dead weights and spring balance is still quite satisfactory for many purposes (note effect of variations in g, p.164).

Hydraulic load cells are simple and robust devices still fairly widely used.

The strain gauge transducer has become the almost universal method of measuring forces and pressures. The technology is very familiar and there are many sources of supply. There are advantages in having the associated amplifier integral with the transducer as the transmitted signals are less liable to corruption, but the operating temperature is then limited.

Force, cyclic

The strain gauge transducer has a limited fatigue life and this renders it unsuitable for the measurement of forces that have a high degree of cyclic variation, though in general there will be no difficulty in coping with moderate variations such as result from torsional oscillations in drive systems.

Piezo-electric transducers are immune from fatigue effects but suffer from the limitation that the piezo-electric crystal that forms the sensing element produces an electrical charge that is proportional to pressure change and which thus requires integration by a charge amplifier. The transducer is unsuitable for steady-state measurements but is effectively the universal choice for in-cylinder and fuel injection pressure measurements. Piezo-electric transducers and signal conditioning instrumentation are in general more expensive than the corresponding strain gauge devices.

Pressure, quasi-static

The liquid manometer has a good deal to recommend it as a device for indicating moderate pressures. It is cheap, effectively self-calibrating and can give a good indication of the degree of unsteadiness arising from such factors as turbulence in a gas flow. Precautions regarding the use of mercury should be observed. The traditional Bourdon gauge with analogue indicator is the automatic choice for many purposes, but see p.333 for a discussion of the sources of error. Regarding strain gauge transducers the comments made above regarding force-measuring devices also apply to pressure measurements.

Pressure, cyclic

The discussion ofthe relative merits of strain gauge and piezo-electric transducers for cyclic force measurement applies equally to their application for pressure measurements.

Capacitative transducers, in which the deflection of a diaphragm under pressure is sensed as change in capacity of an electrical condenser may be the appropriate choice for low pressures, where a strain gauge or piezo-electric sensor may not be sufficiently sensitive.

Position

A simple mechanical linkage may be acceptable provided care is taken to avoid stiction and backlash. Flexible cable type connections are prone to these troubles unless carefully installed and supported. The LVDT (linear variable displacement transducer) is a simple device suitable for sensing linear position and available with total stroke lengths ranging from about 0.25 mm to 0.5 m. Shaft encoders are also used for position indication. Stepper motors are used for the active control and digital indication of position.

Displacement

Inductive transducers are used for 'non-contact' situations, in which displacement is measured as a function of the variable impedance between a sensor coil and a moving conductive target.

The Hall effect transducer is another non-contact device in which the movement of a permanent magnet gives rise to an induced voltage in a sensor made from gold foil. It has the advantage of very small size and is used particularly as an injector needle lift indicator.

Capacitative transducers sense displacements in terms of varying capacity and are useful for measuring very small clearances or changes due to wear.

Acceleration

There are two basic types of accelerometer, based respectively on strain gauge and piezo-electric sensors. Piezo-electric sensors should not be used for frequencies of less than about 3 Hz but tend to be more robust than strain gauge units.

Temperature

Liquid-in-glass thermometers are cheap, simple and easily portable. On the other hand they are fragile, not easy to read and have a relatively large heat capacity and a slow response rate. The interface between the body, solid, liquid or gas of which the temperature is to be measured and the thermometer bulb needs careful consideration. Like all other aspects of temperature measurement there is much guidance available in the literature[2].

Vapour pressure and liquid-in-steel thermometers present the same interface problems, are not as accurate as high-grade liquid-in-glass instruments but are more suitable for the test cell environment and are more easily read.

Liquid-in-glass thermometers are usable over a range from −200°C to +500°C. The other types mentioned above are available for temperatures in the range −20°C to +600°C.

Thermocouples are by far the most widely used temperature measurement transducers. There are many suppliers and a wide range of types. Small-diameter shielded thermocouples are particularly useful for measuring gas and liquid temperatures. The associated instrumentation, essentially a galvanometer with compensated cold junction, is simple and inexpensive. For precise measurements a potentiometer should be used, while small temperature differences may be measured accurately by employing several thermocouples in series. Standardized thermocouple wires are available covering the range from −200°C to +1450°C. Thermocouples applicable to the range of temperatures of concern to the engine tester may be classified as follows:

	Copper–constantan	−250°C to +400°C
Type J	Iron–constantan	−200°C to +850°C
Type K	Chromel–alumel	−200°C to +1100°C

Thermistors (thermally sensitive resistors) have the characteristic that, dependent on the nature of the sensing 'bead' they exhibit a large change in resistance, either increasing or decreasing, over a narrow range of temperature increase. They are particularly useful as safety devices and are generally available in the range from −50°C to +150°C.

Platinum resistance thermometers may be used in the range −250°C to +600°C. They are capable of a higher degree of accuracy than a thermocouple and are particularly useful at low temperatures.

Optical or radiation pyrometers, of which various types exist, are non-contact temperature measuring devices used for such purposes as flame temperature measurement and for specialized research purposes such as the measurement of piston ring surface temperatures by sighting through a hole in the cylinder wall. They are effectively the only means of measuring very high temperatures.

Finally, suction pyrometers, which usually incorporate a thermocouple as temperature sensing device, are the most accurate available means of measuring exhaust gas temperatures, see p.332.

References

I. BS 3403 *Specification for Indicating Tachometer and Speedometer Systems for Industrial, Railway and Marine Use.*
2. BS 1041 Parts 1 to 7 *Temperature Measurement.*

Further reading

BS 6174 *Specification for Differential Pressure Transmitters with Electrical Outputs.*
Hayward, A.T.J. (1977) *Repeatability and Accuracy,* Mechanical Engineering Publications, London.

Appendix II   Standard tests of lubricants

A very large number of test installations are concerned solely with running routine tests of lubricants to ensure that standards are maintained.

In Europe the main authority responsible for such tests is:

The Coordinating European Council for the Development of Performance Tests for Lubricants and Engine Fuels (CEC)
Madou Plaza, 25th Floor
Place Madou 1
B 1030 Brussels, Belgium

A list of current CEC Tests and other publications follows:

L-02-A-78	Oil oxidation and bearing corrosion
L-33-T-82	Biodegradation of two-stroke outboard engine lubricants
L-38-T-87	Valve train scuffing
L-41 -T-93	Evaluation of sludge inhibition qualities of motor oils in a gasoline engine
L-07-A-85	Load carrying capacity of transmission lubricants
L-11-T-72	Coefficient of friction of automatic transmission fluids
L-45-T-93	Viscosity shear stability of transmission lubricants
L-17-A-78	Cam and cylinder wear in diesel engines
L-24-A-78	Engine cleanliness under severe diesel conditions
L-42-A-92	Evaluation of piston deposits and cylinder bore polishing
L-14-A-88	Mechanical shear stability of lubricants containing polymers
L-36-A-90	The measurement of lubricant dynamic viscosity under conditions of high shear (Ravenfield viscometer)
L-37-T-85	Shear stability test for polymer-containing oils
L-39-T-87	Oil/elastomer compatibility test
L-40-T-87	Evaporative loss of lubricating oils
F-03-T-87	Evaluation of gasoline with respect to maintenance of carburettor cleanliness
F-04-A-87	The evaluation of gasoline engine intake system deposits
F-05-T-92	Intake valve cleanliness
M-02-A-78	Internal combustion engine rating procedure
M-07-T-83	The relationship between knock and engine damage
M-08-T-83	Cold weather drivability test procedure
M-09-T-84	Hot weather drivability test procedure

M-l0-T-87   Intake system icing test procedures
M-11 -T-91   Cold weather performance test procedure for diesel vehicles
M-12-T-91   Representative sampling in service of marine crankcase
              lubricants
M-13-T-92   Analysis of marine crankcase lubricants
P-108-79   Manual of CEC reference/standardisation oils for engine/rig
              tests
P-221-93   A guide to the definition of terms relating to the viscosity of
              engine oils
P-222-92   CEC reference fuels manual, 1992, Annual Report
L-18-U-93   Low temperature apparent viscosity
L-12-U-93   Piston cleanliness

Another important source of information describes tests of crankcase, automatic transmission and tractor lubricants laid down by various American
authorities, by Belgian, British and French military authorities and by many
vehicle and engine manufacturers:

Lubricant Specification and Test Manual, PARAMINS, Exxon Chemical
Technology Centre,
PO Box 1
Abingdon OX13 6TT
UK

Index